TRANSACTIONS

OF THE

AMERICAN PHILOSOPHICAL SOCIETY

HELD AT PHILADELPHIA
FOR PROMOTING USEFUL KNOWLEDGE

NEW SERIES—VOLUME 53, PART 5
1963

THE SCENT OF TIME

A STUDY OF THE USE OF FIRE AND INCENSE FOR TIME MEASUREMENT IN ORIENTAL COUNTRIES

SILVIO A. BEDINI

Curator, Division of Mechanical and Civil Engineering, Smithsonian Institution

THE AMERICAN PHILOSOPHICAL SOCIETY
INDEPENDENCE SQUARE
PHILADELPHIA 6

AUGUST, 1963

ACKNOWLEDGMENTS

This writer gratefully acknowledges the considerable assistance he has received from many persons in the collection of information, the translation and interpretation of original sources, and in the assembling of photographs and comparative data of items in private and public collections. Among the many scholars and collectors who have contributed to the preparation of this study are Dr. Teiichi Asahina, of the National Science Museum in Tokyo, who has kindly lent photographs of items in the Museum collection and of the Jikoban at the Nigatsudo at Nara; Professor Derk Bodde of the Department of Graduate Studies at the University of Pennsylvania; Mr. Paul C. Blum of the Charles E. Tuttle Company in Tokyo, Japan; Professor Wolfram Eberhard, of the Department of Sociology at the University of California; Rev. Louis J. Gallagher, S.J., Professor in Chinese Studies at Georgetown University; Professor L. Carrington Goodrich, of the Department of Chinese and Japanese at Columbia University; Miss Elsa Glissman of the New England Institute for Medical Research in Ridgefield, Connecticut; Mr. Dan S. Greenwood of the Harvard College Library at Harvard University; Miss Helen C. Hagar of Derby House, Salem, Massachusetts; Miss Lelia M. Hinkley of the China Institute of America in New York; Mr. Chester W. Howard of McHenry, Illinois; Mr. Edward S. Jones of Los Angeles, California, who is responsible for having initiated this study; Dr. Sang Woon Jeon of Seoul, Korea; Dr. George Kerr of the Honolulu Academy of Arts in Hawaii; Dr. Kojima of the New England Institute for Medical Research in Ridgefield, Connecticut; Mr. Eugene Langston of the Japan Society, Inc., of New York and Tokyo; Mr. Gari Ledyard of the Department of Oriental Languages of the University of California, who translated major sections of the *Hsin Tsuan Hsiang P'u* and the *Hsiang Ch'eng* of Chou Chia-chou, and without whose valuable cooperation this study could never have been written; Dr. S. P. Lehv, M.D., of New York City; Mr. Yorio Miyatake of the Bernice P. Bishop Museum in Honolulu; Mr. Roy Nakajima of the American Embassy in Tokyo; Professor Joseph Needham, Caius & Gonville College of the University of Cambridge; Miss Eleanor Olson, Curator of the Tibetan collection of the Newark (New Jersey) Museum; Dr. Derek J. de Solla Price of the Department of the History of Science and Medicine at Yale University; Mr. Edwin Pugsley of New Haven, Connecticut; Mrs. H. Ivan Rainwater of the Bernice P. Bishop Museum in Honolulu; Professor Edward H. Schafer of the Department of Oriental Languages of the University of California; Rev. Georg Schurhammer, S.J., of the Pontifiche Universita Gregoriana, Vatican City; Father Joseph Sebes, S.J., Professor in Chinese History at Georgetown University; Professor Cyril S. Smith, Professor of Metallurgy at the University of Chicago; Dr. M. K. Starr, Curator of Asiatic Archeology, Chicago Natural History Museum; Dr. William Barclay Stephens of Alameda, California; Dr. Harold P. Stern, Freer Gallery of Art, Washington, D. C.; Mr. Gun'ichi Wada, Secretary of the Shōsōin, Imperial Treasury at Nara, Japan; Dr. F. A. B. Ward of the Science Museum, South Kensington, London; Dr. Alexander Weyman of the Department of Linguistics at the University of California; Professor Kiyoshi Yabuuti of the Research Institute of Humanistic Science at Kyoto University, whose kind interest has solved many problems in interpretation and who has furnished otherwise unobtainable materials; Professor Ruji Yamaguchi of the University of Commerce in Tokyo.

THE SCENT OF TIME

A Study of the Use of Fire and Incense for Time Measurement in Oriental Countries

SILVIO A. BEDINI

As the curling smoke wafts sinuously aloft from a freshly lighted [incense] stick and the first faint whiff strikes the nostrils, a profound symbolism takes effect—even subconsciously. The glowing coal is the spark of life. The smoke signifies the incorporeality and evanescence of spiritual truths, while the fragrance demonstrates the tangible reality and actual penetration of these spiritual truths.[1]

CONTENTS

Introduction... 5
Incense sticks and candles in China..................... 6
The *hsiang-yin* or "incense seal"....................... 6
The Hundred Gradations Incense Seal.................... 7
The Five Watch Incense Seal............................ 12
Wang Ying-lin's *Encyclopedia*.......................... 14
Modern incense-seal timepieces........................ 16
Missionary accounts of Chinese incense timekeepers...... 22
The dragon incense vessel.............................. 26
Incense timekeepers of the Japanese.................... 27
The match-cord.. 27
The candle timepiece.................................. 27
Japanese incense-sticks............................... 27
The geisha timepiece.................................. 28
"Incense seal" timepieces in Japan.................... 29
Captain Golownin's account............................ 29
The *Jo-ko-ban*, or "permanent incense board"........... 29
The *Ji-ko-ban*, or "time measuring incense board"......... 29

The *Mizu-tori*, or spring festival....................... 29
The *Kō-in-za* in the Shōsōin........................... 31
General use of incense timekeepers in Japan............. 33
Father Tçuzu's account of incense timekeepers........... 34
Fraissinet's account.................................. 34
Incense timekeepers in private collections.............. 35
Smelling the time..................................... 37
Incense timekeepers in Korea.......................... 37
Incense timekeepers of Northeast Asia.................. 38
Fire and Incense Clocks in Hawaii..................... 39
Appendices:
   A. Horary systems................................. 40
   B. History of incense............................. 41
   C. The *Mizu-tori* or Japanese spring festival........... 44
Bibliography.. 46
Glossary... 47
Index.. 50

INTRODUCTION

The history of time measurement is filled with many curiosities, none more mysterious and exotic than the examples of the use of fire and incense for timekeeping in Oriental countries.

Until now the so-called fire clocks and incense clocks of China and Japan have evoked only casual comment in the form of brief and passing reference, not only in the histories of science and technology of the Western world, but in the literature and historical accounts of the countries in which they were devised and employed. The lack of recorded information appeared to be indicative of the fact that fire and incense timekeepers were at best no more than novelties, occasionally encountered but not worthy of serious consideration.

It is surprising, therefore, to learn that the use of fire and incense for time measurement has an extensive history, examples having been noted as early as the sixth century A.D. Furthermore, references recently discovered in the works of Oriental writers as well as in the correspondence and the journals of European missionaries to the Far East, indicate that fire and incense clocks served their purpose quite as ade-

quately and as extensively as the sundials and clepsydrae that were the only other timekeepers employed in the epoch preceding the introduction of mechanical timepieces. Consequently, fire and incense clocks merit a place with them in the history of horology, and of Oriental technology.

Briefly, the earliest form of timekeeper used in China appears to have been the sundial, according to a description of one published in the Shu period (third century B.C.). Relics of two individual examples, which were probably made in the Zenkan period (206 B.C.–A.D. 7) were reported by Ryufuku.[2] These were of a primitive form but included indications for the seasons and the compass directions. Their primary function may have been to measure the time only at noon. They may not have been indigenous to China but introduced from another country.

Water clocks appeared to have had a slightly later origin, according to present information, and were probably developed in China by at least A.D. 100–200. An interesting development from these was the sand clock in which water was replaced by sand. This

---

[1] From ancient writings, quoted in Lew Ayres, *Altars of the East* (New York, 1956).

[2] Kiyoshi Yabuuti, "Chugoku No Tokei," *Japanese Journal of the History of Science* 19 (1951): 23–24.

became necessary when water clocks became ineffective in winter owing to the freezing of the water.[3]

In Japan each form of timekeeping device had a later date of appearance. The first recorded water clock is said to have been constructed during the reign of the Emperor Tenchi in A.D. 671. Kiyotsugu[4] reported that Tenchi was an Imperial prince at the time, and that during the reign of the Empress Saimei he produced the first water clock and developed a system of time measurement for the country.

In the Heian period (to A.D. 1140) the Japanese employed a water clock called a *Rokoku* which was constantly maintained in operation by one of two designated officers with twenty assistants.

The fire clocks of China and Japan were simple devices of several basic types which measured time with a fair degree of accuracy and a minimum of equipment.

Perhaps the simplest form was the match cord, which burned like a wick or a fuse.

The graduated candle, with its length marked off into measured intervals, was also well known in China and Japan.

The incense stick or "joss stick" of powdered incense prepared in stick form from hardened paste and marked off along its length into hourly intervals, was the form which probably had the widest distribution in both countries. A variation was the "incense spiral" in which the incense paste was pressed out in the form of a continuous coil and hardened, and marked at regular intervals with the designations for the hours.

Finally, there was the "incense seal" consisting of a continuous trail of powdered incense marked off into equal periods to enable the observer to measure the passage of time by the progress of the burning incense with relative accuracy.

Detailed consideration of the various horary systems of the Chinese and the Japanese is given in Appendix A. A comprehensive summary of the history of incense in the countries of the Far East is included in Appendix B.

INCENSE STICKS AND CANDLES IN CHINA

Although fire may have been employed for time measurement in earlier epochs in China, the first mention of its use for this purpose which has come to notice was in the sixth century. This occurred in the poetic writings of Yü Chien-ku (floruit A.D. 520). Two lines from one of his poems may be translated as follows:

By burning incense [we] know the *k'o* (quarter) of the night, With graduated candle [we] confirm the tally of the watches.

The *k'o* referred to the quarter (equivalent to a period of 14 minutes and 24 seconds) while the "tally of the watches" related to the division of the *keng* or night-watches. (See Appendix A—Horary Systems.)

Although the invention of the candle timepiece has been traditionally ascribed to Alfred the Great of England, it obviously had an earlier history in the Orient. During the period under consideration, Chinese candles were made of beeswax by the dipping method, with a thin, hollow bamboo wick. From the T'ang Dynasty (A.D. 618–907) and later, other materials were also used, such as insect wax and vegetable tallow. The candle was adapted for timetelling by marking off the body into equal segments, labeling each with the appropriate symbol for each time division, and each of which would be consumed within a prescribed period of time.

Incense sticks and graduated candles were both apparently in common use for time measurement during the Sung Dynasty (A.D. 960–1279) for both are found mentioned in the literature of the period. The *Dai Kanwa Jiten* described the graduated candle[5] and later stated that five sticks of incense were burned, one for each of the watches of the night.[6] This would indicate that the period of burning for each stick was from 7 to 11 *k'o*, depending on the season. The incense sticks were probably produced in the manner employed in modern times as described in the foregoing.

THE HSIANG-YIN OR "INCENSE SEAL"

It was during the T'ang Dynasty, that an improved form of incense timepiece came into being, or at least first came to notice in published writings. This was the *hsiang-yin* or "incense seal."

The timepiece was first mentioned in a line of poetry by Fang Kan, whose floruit was in this period, and which was quoted in the *P'ei-wen Yün-fu*:

The yin-hsiang finished [its trail] when [the monk] came out of (i.e., finished) meditation.

The terms *yin-hsiang* and *hsiang*-yin, which are interchangeable, appeared also in the literature of other writers during the next several centuries. Another associated term was *hsiang chuan*, which may be translated as "incense seal-character" and this also occurred in the literature of the period. It was mentioned in the *Hsiang P'u* or "Catalogue of Aromatics (Incense)," a work which was attributed to Hung Ch'u, a writer who worked at the end of the eleventh and beginning of the twelfth centuries. The following passage is taken from his book:

In recent times, persons who esteem oddities have *hsiang chuan* made for them, whose texts conform to the twelve double-hours (*ch'en*); they are divided in a hundred *k'o*, and burn all through the day and night.

---

[3] Joseph Needham, Wang Ling and Derek J. Price, *Heavenly Clockwork*, (Cambridge, 1959), pp. 141, 154 ff.

[4] Hirayama Kiyotsugu, Rekiho Oyobi Jiho.

[5] *Dai Kanwa Jiten* 2: p. 264b.

[6] *Ibid.* 5: p. 963a.

To digress briefly on the subject of terminology, the word *k'o* in the Chinese language literally means a "notch" or "gradation." It came to be the term denoting a period of time equivalent to one quarter of a Chinese "hour." Since the Chinese did not have a word for "hour" in their classical dictionary vocabulary, the term *shih* was used to denote a unit of time equal to a period of two European hours. Thus one *k'o* is equivalent to one half of one European hour, and one quarter of a *shih*. The indicators on the earlier Chinese clepsydrae, and later on the incense clocks, measured time by divisions of the *shih*, each of which was divided into four *k'o*. Consequently the word *k'o* came to be synonymous with the words "clock" or "timepiece."

The word "incense" in Chinese is rendered from the word *hsiang* which literally means "aromatic." Since *hsiang* can mean both "aromatic" and "incense," the context clarifies the meaning to be used.

Although the two Chinese words *yin* and *chuan* are both translated in modern times as "seal," their meaning is actually slightly different. A *yin* is a "seal" in the sense of a stamp or chop employed to impress the outline of a calligraphic character upon a surface. *Chuan*, however, signifies a "seal-character," which is a specific style of calligraphy which was used before the writing reform which occurred in the third and second centuries B.C. Because the style has always remained a favorite for seals and chops, it has been called the "seal form." In their application to the subject of incense timepieces, however, both terms mean the same, i.e., a very stylized pattern resembling "the seal form of the character" (*chuan*), or "the characters that are used on seals" (*yin*). Consequently, the terms "Aromatic (Incense) Seal" or "Aromatic (Incense) Seal-character" are used as names for incense clocks. In the same way, a "notch" (*k'o*) has come to mean clock, and an "aromatic (incense) notch" has become synonymous for incense clock.

Hsiang-yin or "incense seal" may also be given the interpretation of "stamped incense" or "imprinted incense." In such instances, it would be indicative of the fact that a tablet or stick of incense in the form of hard paste was imprinted or engraved with the characters for the time divisions in the traditional Chinese characters used for engraving seals, imprinted at equal intervals to indicate the passage of time with the consumption of the tablet or stick.

The allusion to *hsiang-chuan* or "incense seal-characters" tends to verify this interpretation, and the quotation from Hung Ch'u would easily fit this meaning. In this event, the term *hsiang p'an*, which is frequently encountered in the writings of the twelfth century, and which may be translated as "incense basin," might have been the container in which the incense stick or tablet was burned.

On the other hand, it is just as likely that *hsiang-yin*

or "incense seal" referred specifically to a timepiece consisting of a method of forming a seal character or *chuan* by means of powdered incense, which in turn was applied to and used in a utensil called a *hsiang p'an* or "incense basin." The incense could be formed into a continuous pattern and the resulting trial marked off into equal segments to denote the time periods. Such a device would be consistent with the quotation from the poem by Fang Kan as well as with the description rendered by Hung Ch'u.

The *hsiang p'an* was unquestionably of Buddhist origin and it is probable that the "incense basin" was combined with the famous *Po-shan-lu* of the Han period (202 B.C.–A.D. 140) to become the container for the incense timepiece. It may be noted in this connection that the later Ming bronzes utilized for the burning of incense were called *hsiang lu* and made in a modified form of the earlier bronze sacrificial vessel called a *t'ing*. Meanwhile, a bronze vessel of the Chou period (1030–221 B.C.), which was known as a *p'an* was shaped like a shallow basin and used to hold water in ceremonial rites. The possibility that *hsiang-yin*, *hsiang chuan*, and *hsiang p'an* had a related purpose provides some basis for the belief that in these forms was born the first accurate incense timepiece.

THE HUNDRED GRADATIONS INCENSE SEAL

The earliest reference to a relatively accurate incense timepiece occurred in the eleventh century in the writings of Shen Li, a Chinese prefect with a considerable interest in technology. Shen Li, whose courtesy name or *Tzu* was Li-chih, was famous in his time as a hydraulic engineer, and his work relating to the control of the Yellow River was particularly noteworthy. He died at the age of approximately seventy-one in the reign of Shen-tsung, sometime between the years 1075 and 1086 A.D.

According to Shen Li, an official named Mei-ch'i ("Plumtrees Torrent"), while awaiting promotion, was the inventor of the *pai-k'o hsiang-yin* or "Hundred Gradations Incense Seal" for the purpose of indicating a regular time for sunset and sunrise. The invention was made in A.D. 1073 "in the Kuei-ch'ou Year of the Hsi-ning Reign." The country was beset by a great drought. In the summer and autumn there were only unseasonable rains and the wells and springs dried up and were exhausted, so that the people had little drinking water. It was impracticable to use the customary water clocks or clepsydrae for time measurement.

The incense seal was briefly described by Shen Li in a little work entitled *Hsiang P'u* or "Catalogue of Aromatics (Incense)" which has not survived. A craftsman named Wu Cheng-chung of Kuang-tê (the present province of Chekiang) presumably studied the description and specifications of the incense seal in Shen Li's work and made a practical model of the

Hundred Gradations Incense Seal with improvements of his own. He also produced a recipe for a special incense to be used with it. These he presented to Shen Li in 1074. That dignitary made some trials with the clock and was greatly impressed with its accuracy and worksmanship. So great was his admiration for Wu's talents and his improvements on the original design, that he caused it to be engraved on stone so that he could "pass it on to all those with fond interest in affairs."

Shen Li also described a second type of incense timepiece which he called the *Wu-yeh hsiang-k'o* or the "Five Night Incense Gradations" clock. This was also known by such other variants of the name as "The Five Watch Seal Gradations" and "The Five Night Seal-character Incense Seal" and "The Five Night Incense Seal-character." It was developed for the express purpose of measuring the five night watches.

Although Shen Li's work has not survived as an entity, fragments of his writings were preserved and incorporated into two later works on aromatics and incense, namely, the *Hsin-Tsuan Hsiang-p'u* as well as a rare little work entitled *Hsiang-Sh'eng*.

The *Hsin-Tsuan Hsiang-P'u* or "Newly Compiled Handbook of Aromatics (Incense)" was also known by another title, *Ho-nan Ch'en Shih Hsiang-P'u* or "Handbook of Incense by [a member of] the Ch'en Family of Honan." It was compiled by Ch'en Ching of Honan probably between A.D. 1275 and 1322. There is a possibility, however, that the work was completed as early as the middle or latter part of the twelfth century.

Ch'en Ching collected into a single work what earlier writers had produced on the subject of aromatics and incense. He assembled available fragments of Shen Li's writings on the subject and he presumably had access to the stone inscription of 1074. Other texts included "The Catalogue of Incense" or *Hsiang P'u* by Hung Ch'u of the first half of the twelfth century, "The History of Aromatics" by a writer named Yen, and a "Catalogue of Incense" by another named Yeh, which was written in 1151, as well as the work of Hsiung P'eng-li. Nothing is known about the writer named Yen. Yeh, however, was a customs inspector at Ch'üan-chou, a southern seaport, where he may have become familiar with incense being imported from southeastern Asia. Hsiung P'eng-li was described as "the Futile Fisherman of Lake P'eng-li" near Yü-chang, who wrote a preface to his work in the "Orchid Autumn (seventh lunar month) of the jen-hsü year of the Chih-chih Reign" or in the year A.D. 1322. Of the eleven sources listed by Ch'en Ching only the Catalogue of Incense by Hung Ch'u has survived.

The clue to the date of Ch'en Ching's compilation may lie in the inclusion of one of his sources, the work entitled "Ch'ien chai Aromatic (Incense) Catalogue

Supplement" which may date as late as the end of the thirteenth century. Ch'ien chai is the pen name of two writers who lived at the end of the Sung and the beginning of the Yüan periods, or near the end of the thirteenth century. On the other hand, Ch'ien chai may be the pseudonym of some earlier writer. It is apparent from the text that Shen Li's work survived in fragmentary form.

Ch'en Ching compiled into a single work all that remained of what earlier authors had written on the subject of aromatics and incense. The *Hsiang-P'u* was later included as part of the *Shih-Yüan Ts'ung-Shu* as well as the important work entitled *Hsiang-Sh'eng* or "Comprehensive Account of Aromatics (Incense)" which was produced between 1618 and 1641 by Chou Chia-chou (floruit 1580–1650).

Shen Li's account of the incense seals occurs in *chuan* II or the second chapter of the *Hsin Tsuan Hsiang-P'u*, the first paragraph of which is entitled "The Five Night Incense Gradation-Clock (carved in stone at Hsüan-chou)." The first part of the text is devoted to a brief summary of the employment of water clocks to that time, as follows:

[Time-keeping instruments] with a hollow pitcher as clepsydra and a piece of floating wood as an indicator-rod, have been esteemed since the time of the Yu-hsiung Clansman.[7] That usage has been followed from the three periods of antiquity, through the two Han Dynasties,[8] down to the present day.[9] Although the construction of these [clepsydrae] has at times been clever, and at other times clumsy, there has been no reason to change from this system. At the beginning of our Dynasty (the Sung period, 907–1279 A.D.) we acquired the water steelyard of the T'ang Court (608–906 A.D.). Its functioning has been delicate and ingenious. It corresponds very closely to the steelyard clepsydrae of Tu Mu at Hsüan-chou and Jun-chou. Later, in our Dynasty (Yuan or Mongol period, 1322 or earlier), when Yen Su of the Dragon Chart Library[10] was Magistrate of Tzŭ-chou, he made the Lotus Flower Clepsydra, and presented it to His Highness. Recently, in addition, Wu Seng-jui has newly created the steelyard clepsydrae of Hang-chou and Hu-chou, but these examples are both inaccurate and imprecise. At the beginning of the Wu-tzu Year of the Ch'ing-li Reign (1048 A.D.), it was desired to set the Court ceremonials in new order. In the Twelfth Month, the Court Recorders withdrew and made an announcement permitting all officials to look over the new steelyard clepsydra in the Court Hall. Consequently, they obtained a detailed look at it, and committed it silently to memory. It was only then that they knew that neither the ancient nor the modern (T'ang) systems had been minutely thought out or probed to their deepest. Pre-

---

[7] The Yu-hsiung Clansman is another name for Huang-ti, a mythical emperor, who, according to the *Kuo Shih Chih* (probably *ca.* A.D. 1082), invented the clepsydra by watching dripping of water.

[8] The two Han Dynasties, called by some the Western and Eastern, and by others, the Former and Later Dynasties, dated from 206 B.C. to A.D. 9 and from A.D. 25 to A.D. 220 respectively. They were separated by the Wang Mang Interregnum.

[9] Eleventh century A.D.

[10] The Dragon Chart Library was a repository for storing Imperial writings and book collections.

sumably, if you make the water case for the second steel-yard small, it causes either slowness or celerity in the dripping from the clepsydra. Have not the difficulties of all times past been grandly corrected by the research of our Dynasty? Once I brought forth my own stupid shortcomings and presumed to develop a method in imitation [of the above discovery] which I applied in the Drum and Cornet Towers of Wu-chou and Mu-chou.[11]

The foregoing was of particular significance since it provided a picture of timetelling methods of the period supplementing the data which have already been collected on the subject in recent times. The text of *chuan* II continues with Shen Li's account of the invention of the incense seal:

In the Kuei-ch'ou Year of the Hsi-ning Reign (1073 A.D.), there was a great drought. In the summer and autumn, [there were only] unseasonable rains, and the wells and springs were dried up and exhausted. The people were hard pressed for drinking water. It was at this time that Mei-ch'i ("Plumtrees Torrent"),[12] an official awaiting promotion, first made the Hundred Gradations Incense Seal, in order to regulate sunset and sunrise. In addition, he set up the Five Night Incense Seal, as follows[13]

The Hundred Gradations Incense Seal is made of hard wood. Mountain Pear (*Pirus serotina var. culta*) is best, with Camphor (*Cinnamomum camphora*) and the Nanmu Tree (*Machilus nanmu*) next, in that order. It [the Seal] is 1.2 inches thick.[14] The outer diameter is 1 foot 1 inch. The diameter of the center part is 1 inch and no more. Take the part with the pattern and divide it into twelve sectors of zig zag lines. The pattern is perpendicular [to the sector radii], with 21 paths lying over [each other]. All paths are .15 inches in breadth. Bring the top part to a point. The depth is likewise like this. Each gradation (*k'o*) is 2.4 inches in length. One hundred gradations has a total length of 240 inches. Each HOUR (*shih*) is almost 2 feet in length, and the total [of the twelve HOURS] is 240 inches. Each HOUR contains eight and one-third gradations.[15]

In the narrow part close to the center, six of the ring [segments] connect with the adjoining [ring segments, making six pairs:] PIG and RAT, OX and TIGER, RABBIT and DRAGON, SNAKE and HORSE, SHEEP

FIG. 1.　One hundred gradation incense seal.
From *Hsin Tsuan Hsiang-P'u.*

and MONKEY, and COCK and DOG. *Yin* exhausts itself to reach *Yang.*

When [the flame] is at the end of DOG, it enters PIG, because the six long ring [segments] each connect with each other on the outer rim.

There are six *Yang* HOURS.[16] They all go forward, from the small into the large, from the imperceptible to the manifest.

At the long parts facing the outer rim, six of the ring [segments] likewise connect with the adjoining [ring segments], making six other pairs: RAT and OX, TIGER and RABBIT, DRAGON and SNAKE, HORSE and SHEEP, MONKEY and COCK, and DOG and PIG. The *Yang* culminates in order to enter *Yin.*

When [the flame] is at the end of PIG, it reaches RAT, because the six [short] ring [segments] each connect with each other in the center.

There are six *Yin* HOURS. They all go backward, from the large into the small. *Yin* presides over the extinction [of the flame].

There is no stopping point at all. It is similar to a ring, which has no end. Each time one starts the fire, one

---

[11] The Drum and Cornet Towers were probably watchtowers from which alarms were sounded in the event of trouble.

[12] Mei-ch'i or "Plumtrees Torrent" appears to be a pen-name, the identity of which cannot be ascertained. It was a pen-name used by two scholars in the twelfth and early thirteenth centuries, but no person has been found so named for the period referred to, i.e., A.D. 1073.

[13] Although the final sentence in the paragraph mentions the Five Nights Incense Gradation, the author proceeded to describe the Hundred Gradations Incense Seal, followed subsequently by a description of the Five Watch Incense Seal. Presumably the *Five Nights* was intended to read *Five Watch*, etc.

[14] The *ts'un* (inch) is equivalent to 1/10 of a *ch'ih* (foot).

[15] If 2.4 inches are multiplied by the decimal 8.333, the variation is very slight. Twelve hours contain exactly 100 *k'o* or gradations. One *shih* (double-hour) is equal to 8 and 1/3 *k'o*. According to the encyclopedic Chinese-Japanese dictionary, the *Dai Kanwa Jiten,* "In ancient times a day and a night were divided into 100 *k'o*. At the vernal and autumnal equinoxes, there were 50 *k'o* each for the day and for the night. At the Winter Solstice, there were 40 *k'o* for the day and 60 *k'o* for the night, while at the Summer Solstice it was vice versa." This definition referred to *k'o* as is used in clepsydrae.

[16] The *Yin* and the *Yang* hours are indicated on the perimeter of the diagram of the Hundred Gradations Incense Seal. The *Yin* hours are those for which the flame-track or incense-trail narrows or reduces from the outside inward. This is in harmony with the concept of *yin,* which is negative, female, weak, waning, decreasing. *Yin* never wholly disappears; when it reaches the end of its process it becomes *yang* and begins the opposite course. In traditional Chinese thought, even numbers are considered *yin,* and the *yin hours* on this clock are the 2nd, 4th, 6th, 8th, 10th and 12th. The *Yang* hours are those for which the flame-track widens or expands from the inside outward. *Yang* is positive, male, strong, waxing, increasing. The odd-numbered hours are the *Yang hours,* namely, the 1st, 3rd, 5th, 7th, 9th and 11th. Another meaning for *Yang hours* is "sunlight hours" and for "*Yin hours*" it is "night time hours."

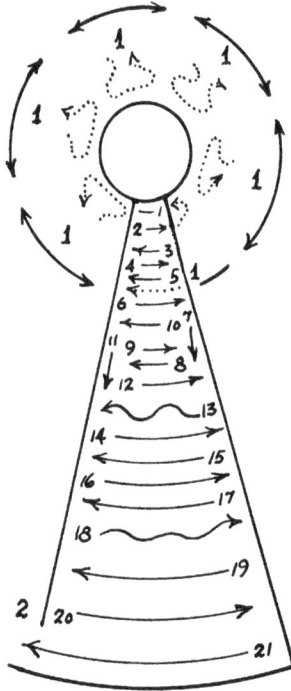

FIG. 2. Overlay on hundred gradations incense seal.

goes by the [appropriate] HOUR. Generally, one starts at the Top 'o the HORSE [noon].[17]

The third path close to the center is the correct [place].[18]

Some start [the fire] at sunrise.

[For example], one looks at the calendar [and finds that] sunrise is in RABBIT, [so he starts the fire] at Top 'o the RABBIT [or] several gradations [to one side or the other as appropriate].

We set neither an end point nor a place for starting the fire.

The foregoing translation was rendered from the *Hsin Tsuan Hsiang-P'u* and amplified by comparison with the related passages in the *Hsiang Ch'eng*. Shen Li's instructions for the construction and operation of the *pai-k'o hsiang-yin* were quite specific and detailed. In the translation the Chinese word *Shih*

[17] "Top 'o the Horse" is a rather fanciful ranslation for the Chinese *wu-cheng*, which means literally "the HORSE upright" and refers to the middle point of the HOUR. The *wu* or HORSE HOUR lasts from 11:00 A.M. to 1:00 P.M. and the "HORSE upright" consequently means high noon.

[18] This Chinese note is puzzling, for if one wished to start the clock at noon, it would seem that it would be necessary to start it in the middle of the winding path that traverses the HORSE sector. However, the note seems to indicate that the fire should be started in the third path from the center (as indicated in the overlay) which would appear to make the clock "slow."

has been rendered as HOUR capitalized to distinguish it from the European hour of sixty minutes. Each *Shih* is equivalent to 2 European hours, and twelve *Shih* elapse in the course of a complete day and night interval. Each *Shih* is named for one of the characters in the list of "Twelve Branches" and is symbolized by one of the twelve animals. The names of the animals have been used to translate these cyclical signs and capitalized to indicate that each represents an HOUR instead of an hour. The Chinese *k'o* may be rendered as "gradation" although its definition is technically "notch." It represents an interval of 14 4/10 minutes in European time, inasmuch as each *Shih* or double-hour includes $8\frac{1}{3}$ *k'o*.

The "Hundred Gradations Incense Seal" illustrated in the *Hsin Tsuan Hsiang-P'u* is reproduced in figure 1. It is quite probable that this was taken from the stone engraving of Shen Li. The "part with the pattern" to be divided into twelve sectors of zigzag lines is understood to mean the area of the total circle exclusive of the center ring, as indicated in figure 2.

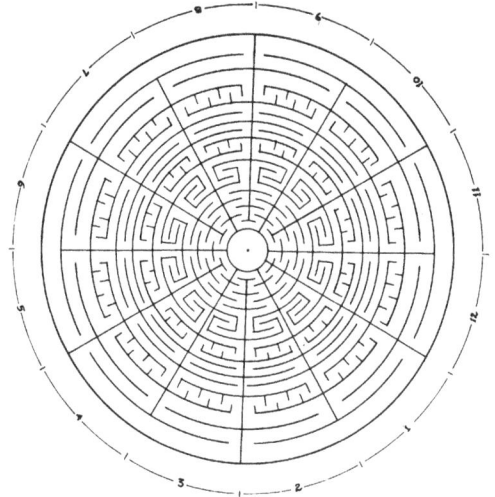

FIG. 3. Diagram of hundred gradations incense seal with explanatory notations added.

KEY *To Diagram of the "Hundred Notch Seal-Character Aromatic"*

1.	*Wu*	Horse	11:00 A.M.– 1:00 P.M.	(Yang)
2.	*Wei*	Sheep	1:00 P.M.– 3:00 P.M.	(Yin)
3.	*Shen*	Monkey	3:00 P.M.– 5:00 P.M.	(Yang)
4.	*Yu*	Cock	5:00 P.M.– 7:00 P.M.	(Yin)
5.	*Hsü*	Dog	7:00 P.M.– 9:00 P.M.	(Yang)
6.	*Hai*	Boar or Pig	9:00 P.M.–11:00 P.M.	(Yin)
7.	*Tzu*	Rat	11:00 P.M.– 1:00 A.M.	(Yang)
8.	*Ch'ou*	Ox	1:00 A.M.– 3:00 A.M.	(Yin)
9.	*Yin*	Tiger	3:00 A.M.– 5:00 A.M.	(Yang)
10.	*Mao*	Hare or Rabbit	5:00 A.M.– 7:00 A.M.	(Yin)
11.	*Ch'en*	Dragon	7:00 A.M.– 9:00 A.M.	(Yang)
12.	*SSu*	Snake	9:00 A.M.–11:00 A.M.	(Yin)

In the next line, rendered as "the pattern is perpendicular," the original text stated that "the pattern is criss-cross" in respect to the twelve sectors which, when drawn, define twelve areas crossing back and forth across them. Figure 3 is a reproduction of the "Hundred Gradations Incense Seal" with descriptive details added to clarify the foregoing text.

Shen Li then proceeded to provide a suitable recipe for the preparation of the incense to be used with the *pai-k'o hsiang-yin*, presumably as originated by Wu Cheng-chung:

In the case of the Hundred Gradations Incense Seal-Character, if one uses the ordinary incense it will not burn evenly. Now we use the two flavors "Wild Thyme" and "Pine Ball." Mix them together evenly and store them in a new earthenware vessel, and then [they may be] applied and used. The "Wild Thyme" is the leaf of the Jen Tree (*Perilla ocimoides*). Pick the leaves just before the middle of autumn, lay them out in the sun to dry, and then make them into powder. Ten ounces for each batch. The 'Pine Ball' is the withered pine flower.[19] Gather them at the end of autumn, taking the ones that have fallen by themselves. Lay them out in the sun to dry, cut away the center, and make the remainder into powder. For each batch use eight ounces.

Formerly, when I compiled the preface for the *Hsiang P'u* (Catalogue of Incense), my account of the Hundred Gradations Incense was not very detailed. Wu Cheng-chung of Kuang-te fashioned his Seal-character Gradation as well as the recipe for the incense. On being presented with it, I tried it out, and I have found that it is quite minutely thought out and considered. Were it not for his refined talents and subtle thought, how could we have been able to attain to this? Consequently, I have engraved it on stone so as to pass it on to all those with fond interests in affairs.

The Second Day of the Second Month of the Chia-yin Year of the Hsi-ning Reign (1074). Right Censor, Grandee, and Magistrate of Hsüan-ch'eng, Shen Li.

The remainder of the volume is devoted to data and recipes for the different types of incense to be used in the incense timepieces. Of the thirteen separate recipes, two examples have been selected:

The Hundred Gradations Seal Incense

Chien-hsiang (unidentified).................... 2 oz.
T'an-hsiang (Sandalwood, *Santalum album*)..... 2 oz.[20]
Ch'en-hsiang ("Sinking Aromatic," *Aquilaria agallocha*)................................ 2 oz.[21]
Huang-shu-hsiang (*Aquilaria agallocha*)........ 2 oz.
Ling-ling-hsiang (*Ocimum basilicum*)........... 2 oz.
Huo-hsiang (*Lophantus rugosus*)................ 2 oz.
T'u-ts'ao-hsiang ("Dirt Grass Aromatic," unidentified), with the dirt removed............ ½ oz.

Mao-hsiang ("Reed Aromatic," Geranium Grass, *Andropogon Schoenanthus*)................... 2 oz.
P'en-hsiao ("Bowl Flux")[22].................... ½ oz.
Ting-hsiang ("Cloves, *Eugenia aromatica*)........ ½ oz.
Chih-chia-hsiang (Unidentified).............. 0.75 oz.
Lung-nao ("Dragon Brain," Borneo Camphor, *Dryobalanops aromatica*)................... a little.
Take the above ingredients and combine them into a powder. Burn it according to the usual method."

This recipe was selected for inclusion in this study since its name implies that it was specifically to be used with the Hundred Gradations Incense Seal, and presumably of the same period. Another typical recipe from the collection was

The Ting-chou Public Storehouse Seal Incense

Chien-hsiang (unidentified)..................... 1 oz.
T'an-hsiang (Sandalwood, *Santalum album*)....... 1 oz.
Ling-ling-hsiang ("Misty Tumulus Aromatic"),[23] *Ocimum basilicum*)........................... 1 oz.
Huo-hsiang (*Lophantus rugosus*)................. 1 oz.
Kan-sung ("Sweet Pine" Spikenard, *Nardostachys jatamansi*)................................. 1 oz.
Ta-huang ("Big Yellow," Rhubarb, Rheum sp.)... 1 oz.
Mao-hsiang ("Reed Aromatic," Geranium Grass, *Andropogon Schoenanthus*), very finely powdered. By soaking it in water, by drying it in the sun, and then roasting it with a slow fire, you can make the color yellow.............. ½ oz.

Take the above ingredients and pestle them through a sieve, making powder, and then use it according to the usual method. Whenever one makes Seals or Seal-characters, the powder must be mixed with a little almond powder, the incense will then not raise dust. Everyone follows this method after they have had some aromatic (incense) blow away on them.

Another elaborate incense clock was included in both the *Hsiang Ch'eng* and the *Hsin Tsuan Hsiang-P'u* and is reproduced in figure 4 from the latter volume. The drawing was entitled "Picture of the Greatly Elaborated Incense Seal Character" and according to the caption

Tsou Hsiang-hun was presented with this diagram. Hsiang-hun, whose adult name was Hsi-lung, and whose courtesy name was Shao-nan, was a native of Yü-chang (the present province of Kiangsi). His post was in Tzŭ-li [Prefecture] of Yü-li. He loved the old "Broad Elegancies." He was excellent on "The Book of Odes," and capable in literature. He was especially excellent on "The Book of Changes." He was held in high esteem by many worthy gentlemen and grandees.

First day of the Good Month (October) in the Second year of the T'ien-li Reign (1329 A.D.). Written by the Retired Gentleman of the Central Studio.

---

[19] It should be noted that the *flower* of the pine tree is meant, not the *cone*.

[20] Bernard Read identifies *T'an-hsiang* as "Red Sanders Wood" (*Pterocarpus santalinus*). However, Professor Edward H. Schafer has recently established its real identity as Sandalwood.

[21] Both *Ch'en hsiang* and *Huang-shu-hsiang* are derived from the same tree, *Aquilaria agallocha*. The *Ch'en hsiang* is dark while the *Huang-shu-hsiang* is light, withered and not very hard.

[22] According to the *Dai Kanwa Jiten* (Vol. 8, No. 22959.30 and Vol. 6, No. 14428.21, under *p'u hsiao*, "It is produced in salty regions, and resembles table salt. When boiled in water, it becomes crystalline. The color is a pale yellow, and the texture is coarse. In melted form it is used to soften the hides of oxen and horses." It is also called "Bowl flux" because after heating, if it is placed in a bowl, it is said to congeal.

[23] This aromatic is named after a range of mountains in southern Hunan.

FIG. 4. "Greatly Elaborated Incense Seal" from *Hsin Tsuan Hsiang-P'u.*

An additional cautionary note appeared under the "Picture of the Greatly Elaborated Incense Seal-character," and the reader was advised that

Whenever aromatic powders for Seals or Seal-characters are mixed, do not use Chien [aromatic], or "Nipple [Aromatic]," *Pistacia khinjuk,* or Chiang-chen Aromatic (*Rutaceae Acronychia laurifolia*), for they are oily, liquid and bubbly, and cause the fire not to burn. The various recipes are listed in detail in the previous chapter.

THE FIVE WATCH INCENSE SEAL

The editor then preceeded to present a condensed description of the *Wu-yêh hsiang-k'o* or "The Five Watch Incense Seal," which had also been engraved on stone by Shen Li:

The First Seal is the longest. It has a single use: from after [the festival day of] "Slight Snow," through "Heavy Snow" and "Winter Solstice," to after [the festival day of] "Slight chill." Following it are the four Seals A, B, C and D, which have a double, use, and which have two gradations [totals each].

The Middle Seal is the most average. It has a single use: from after [the festival day of] "Excited Insects" to after [the festival day of] "Vernal Equinox," and likewise

UNIFIED DATA FOR THE "FIVE NIGHT SEAL–CHARACTER AROMATIC"

1. Seal	2. Number of notches (*k'o*)	3. Diameter	4. Length	5. Dates covered, according to the text	6. Dates covered, according to emendations made in the text
First Seal	60	3.3″	2′7.5″	*Dec 2*–Jan 9	Dec 3–Jan 9
Seal A	59 58	3.2″	2′7″	Jan 10–Jan 23, Nov 21–Dec 3 Jan 24–Feb 1, Nov 11–Nov 20	Jan 10–Jan 23, Nov 21–Dec 2 Jan 24–Feb 1, Nov 11–Nov 20
Seal B	57 56	3.2″	2′6″	Feb 2–Feb 9, *Jan 31–Feb 8* Feb 10–*Feb 17, Oct 19*–Nov 2	Feb 2–Feb 9, Nov 3–Nov 10 Feb 10–Feb 16, Oct 27–Nov 2
Seal C	55 54	3.2″	2′5″	*Feb 16*–Feb 22, Oct 21–Oct 26 Feb 23–*Mar 1*, Oct 14–Oct 20	Feb 17–Feb 22, Oct 21–Oct 26 Feb 23–Feb 28, Oct 14–Oct 20
Seal D	53 52	3″	2′4″	Mar 1–Mar 5, Oct 7–Oct 13 Mar 6–Mar 11, Oct 1–Oct 6	Mar 1–Mar 5, Oct 7–Oct 13 Mar 6–Mar 11, Oct 1–Oct 6
Seal E	51	2.9″	2′3″	Mar 12–Mar 17, Sep 26–*Oct 1*	Mar 12–Mar 17, Sep 26–Sep 30
Middle Seal	50	2.8″	2′2.5″	Mar 18–Mar 23, Sep 21–Sep 25	Mar 18–Mar 23, Sep 21–Sep 25
Seal F	49	2.8″	2′2″	Mar 24–Mar 28, Sep 15–Sep 20	Mar 24–Mar 28, Sep 15–Sep 20
Seal G	48 47	2.7″	2′1.5″	Mar 29–*Apr 1*, Sep 9–Sep 14 Apr 4–Apr 11, Sep 3–Sep 8	Mar 29–Apr 3, Sep 9–Sep 14 Apr 4–Apr 11, Sep 3–Sep 8
Seal H	46 45	2.6″	2′.5″	Apr 12–Apr 17, Aug 27–Sep 2 Apr 18–Apr 23, Aug 19–Aug 26	Apr 12–Apr 17, Aug 27–Sep 2 Apr 18–Apr 23, Aug 19–Aug 26
Seal I	44 43	2.5″	1′9.5″	Apr 24–Apr 30, Aug 12–Aug 18 May 1–May 8, Aug 4–Aug 11	Apr 24–Apr 30, Aug 12–Aug 18 May 1–May 8, Aug 4–Aug 11
Seal J	42 41	2.4″	1′8.5″	May 9–May 18, Jul 25–Aug 3 *May 20*–Jun 1, Jul 11–Jul 24	May 9–May 18, Jul 25–Aug 3 May 19–Jun 1, Jul 11–Jul 24
Last Seal	40	2.3″	1′7.5″	*Jun 3*–Jul 10	Jun 2–Jul 10

FIG. 5. Table of unified data for the five watch incense seal.

[before and after] the "Autumnal Equinox." Preceding and following it are the [two] seals E and F respectively. They have a double use and a single [gradation total].

The Last Seal is the shortest. It has a single use: from before [the festival day of] "Grain in Ear," through "Summer Solstice," to after [the festival day of] "Slight Heat." Preceding it are the four seals G, H, I and J, which have a double use, and which have two gradations [totals each].

The First Seal is the longest because it has the maximum number of *k'o* or gradations, a total of 60. It is used only during the period of the year when the nights are approximately 60 *k'o* long. These are the nights from after the festival day of "Slight Snow" or November 22, to after that of the festival day of "Slight Chill," or January 6. In actuality, these dates are approximate only, as the "Table of Unified Data For the Five Night Seal Character Incense Clock" included as figure 5 clarifies. The four Seals A, B, C, and D have a double use in that they can be used two times during the year.[24] As an example, Seal A can be used both before and after the Winter Solstice. Seals A, B, C, and D also have two sets of lengths. Seal A can be either 59 or 58 *k'o*, for instance. The text is not explicit regarding how this variability is achieved, but the diagrams make it apparent that Seals A, B, C, and D each have a curlicue at the center into which the flame track presumably is allowed to continue to allow for the greater number of *k'o*, as 59 in the case of Seal A. If, on the other hand, the curlicue is closed off, it allows for the lesser number, or 58 *k'o*.

The Middle Seal is the most average because the total *k'o*, which is 50, is the mean between 40 and 60 *k'o*, the totals of the First and Last Seals, respectively. Its single use is during the periods when the nights are approximately 50 *k'o* long, or just before and after the two equinoxes. Seals E and F can be used twice during the year, as shown on the Table. Unlike Seals A, B, C, and D, Seals E and F have only one *k'o* total respectively: Seal E has 51 *k'o* and Seal F has 49. Neither Seals E or F have a curlicue in the center, bolstering the supposition concerning the function of the curlicue in Seals A, B, C, and D.

The Last Seal is the shortest because it has the minimum number of *k'o*, or a total of 40. Its single period of use is during the time when the nights are approximately 40 *k'o* in length, as indicated on the table in figure 5. Seals G, H, I, and J function similarly to Seals A, B, C, and D.

In the text following the brief summation of the function of "The Five Watch Seal Character" is a series of 13 drawings with captions which serve to explain the use of this form of incense clock, and these are reproduced in Figures 6, 7 and 8.[25]

---

[24] These are the first four of the "Ten Celestial Stems," namely, *chia, i, ping* and *ting. Wu* and *chi* are the fifth and sixth Stems, and the last four Stems are *keng, hsin, jen* and *kuei.*

[25] The thirteen drawings of the variations of the Five Watch Seal Gradation included here are reproduced from the *Hsin*

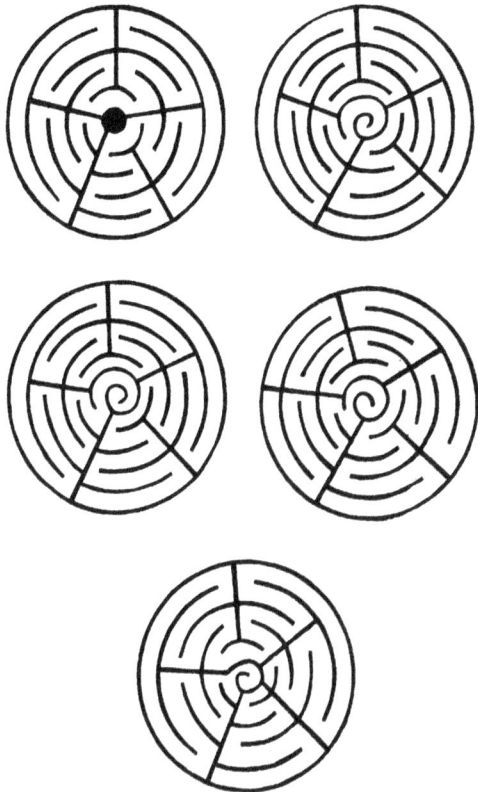

FIG. 6.   *Five Night Incense Seal:* First seal—Seals A, B, C, D, E.

The instructions embodied in the captions have been compiled and arranged in figure 5 in the Table of Unified Data For the Five Night Seal-Character Incense Timepiece" for facility of comparison. The Seals are listed by name as they appeared in the text, followed by the number of *k'o* or notches for each as designated. The diameter of the character is given in each instance, in addition to a measurement of the length of the flame track or incense path, followed by the dates that each is to be used.

In nine instances the dates have been found to be in conflict with other data in the column. Some of these conflicts were due to an obvious corruption in the text, and in others to a disparity between the approximate dates provided in Appendix A of Mathews' *Dictionary of Chinese Dates*, and the dates derived by reckoning the specifications of the text. In each

Tsuan Hsiang-P'u, in which they appeared without captions. The same designs were illustrated in the twenty-second chüan of the *Hsiang Ch'eng* with descriptive captions, but the drawings were of much poorer quality. The edition of the *Hsin Tsuan Hsiang-P'u* which has been used for this study is from the *Shih-yüan ts'ung shu,* published by *Chang Chün-heng* in 1914.

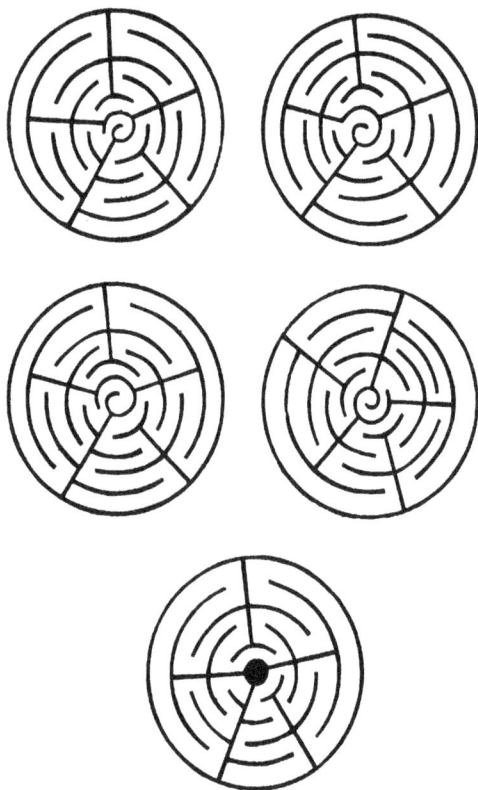

FIG. 7.   *Five Night Incense Seal:* Middle seal—
Seals F, G, H, I, and J.

instance of conflict the dates have been amended to fit the scheme of the Seals. The dates as amended are tabulated in the final column, and these are shown to be consistent with each other.

Figure 9 is a chart which has been plotted for the duration of use for the various Seals against an approximate elliptical curve showing the length of the night as it varies throughout the year. The dotted lines represent the dates for which the Chinese text

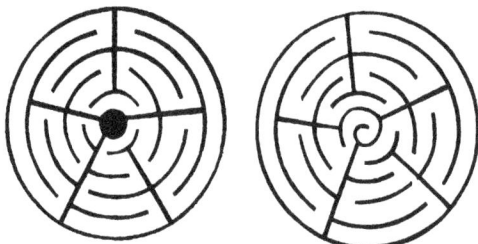

FIG. 8.   *Five Night Incense Seal:* Last seal.

is in conflict but which has been amended as shown in the table in figure 5.

At the end of the twenty-second *chüan* of the *Hsiang Ch'eng* there are two seal characters for incense clocks which are not found in the *Hsin Tsuan Hsiang P'u*, and for which no explanation is given in the text. The first example, reproduced in figure 10*a*, is the Chinese character *shou*[26] in a very abbreviated and stylized form. The second is the character *fu* and reproduced in figure 10b, also greatly stylized.

WANG YING-LIN'S ENCYCLOPEDIA

Verification of the employment of "incense seals" for timekeeping occurred in another early Chinese reference work entitled *Hsiao Hsüeh Kan Chu* issued in ten *chüan* or parts. This encyclopedia was compiled by Wang Ying-lin (A.D. 1223–1296) in about the year A.D. 1270 although it was not published until 1299, a few years before his death. Wang Ying-lin was a prolific writer and this was but one of many works from his pen. In the first *chüan* of this comprehensive work he quoted an earlier writer named Hsüeh-Chi-Hsüan of the first half of the twelfth century as having stated that

In these times (i.e., twelfth century) timekeeping devices (*kuei lou*) are of four different types. There are the bronze vessels [clepsydrae] (*t'ung hu*), the [burning] incense seal-characters (*hsiang chuan*), the sundial (*kuei piao*), and the revolving and snapping springs (*kun t'an*). . . .

Professor Kiyoshi Yabuuti[27] refers to the same passage from the *Hsiao Hsüeh Kan Chu,* adding that

[The] *hsiang chuan* is a clock which measures time by means of a joss stick. This clock was called a *yen chuan* in a work entitled "*Lou K'ê Ching*" by an unknown author which was published probably in the 5th to the 6th Centuries, and it is clear that such a clock or timepiece was used since that time. These clocks were employed for private or personal use in order to measure time at night, but they were too inaccurate to be used for official or public purposes.

Although the phrase *hsiang chuan* in the passage from Wang Ying-lin's work has been translated as "incense stick" in other recent studies on the subject of Chinese horology, it is more accurately rendered as "incense seal-characters" or "incense seal-character timepiece," inasmuch as the term for incense stick is *hsiang-pang* or *pang-hsiang*. Wang Ying-lin consequently referred to the same form of incense timepiece already noted in the twelfth century and which may be identified with the *hsiang-yin* noted in the ninth century. On the basis of the reference to the

---

[26] The Chinese character *shou* represents "long life," a thought frequently represented in the decorative art of China. The actual character evolved into many stylized variations, numbering more than one hundred.

[27] Kiyoshi Yabuuti, Chugoku No Tokei, *Japanese Journal of the History of Science* 19 (1951): 23.

CHART FOR THE FIVE NIGHT SEAL CHARACTER AROMATIC

Solar Period	Seal	Number of Notches	Days Used	Night Time
Feb. 5 (Beginning of Spring)	Seal B	57 k'o	8	
	Seal B	56 k'o	(7)	
	Seal C	55 k'o	(6)	
Feb. 19 (Rain Waters)	Seal C	54 k'o	(6)	
	Seal D	53 k'o	(5)	
March 5 (Excited Insects)	Seal D	52 k'o	6	
	Seal E	51 k'o	6	
	Middle Seal	50 k'o	6	
March 20 (Vernal Equinox)	Seal F	49 k'o	5	
	Seal G	48 k'o	(6)	
	Seal G	47 k'o	(8)	
April 5 (Clear Brightness)	Seal H	46 k'o	6	
	Seal H	45 k'o	6	
April 20 (Grain Rains)	Seal I	44 k'o	7	
	Seal I	43 k'o	8	
May 5 (Beginning of Summer)	Seal J	42 k'o	10	
May 21 (Small filling [of Grain])	Seal J	41 k'o	14	
June 6 (Grain in Ear)	Last Seal	40 k'o	39	
June 21 (Summer Solstice)				
July 7 (Slight Heat)	Seal J	41 k'o	14	
July 23 (Great Heat)	Seal J	42 k'o	10	
	Seal I	43 k'o	8	
Aug. 7 (Beginning of Autumn)	Seal I	44 k'o	7	
	Seal H	45 k'o	8	
Aug. 23 (Abiding Heat)	Seal H	46 k'o	7	
	Seal G	47 k'o	6	
Sept. 8 (White Dew)	Seal G	48 k'o	6	
	Seal F	49 k'o	6	
	Middle Seal	50 k'o	5	
Sept. 23 (Autumnal Equinox)	Seal E	51 k'o	(5)	
	Seal D	52 k'o	(6)	
Oct. 8 (Cold Dew)	Seal D	53 k'o	7	
	Seal C	54 k'o	7	
Oct. 23 (Frost Falls)	Seal C	55 k'o	6	
	Seal B	56 k'o	(7)	
Nov. 7 (Beginning of Winter)	Seal B	57 k'o	(8)	
	Seal A	58 k'o	10	
Nov. 22 (Slight Snow)	Seal A	59 k'o	(12)	
Dec. 7 (Heavy Snow)	First Seal		(38)	
Dec. 21 (Winter Solstice)				
Jan. 6 (Slight Chill)	Seal A 59 k'o		13	
Jan. 21 (Great Chill)	Seal A 58 k'o		9	
	Seal B	57 k'o	8	

FIG. 9. Chart showing duration of *Five Night Incense Seal.*

FIG. 10a. *Shou* Seal
from *Hsiang Ch'eng.*

FIG. 10b. *Fu* Seal,
from *Hsiang Ch'eng.*

Lou K'ê Ching ("The Book of the Clepsydra") noted by Professor Yabuuti, the incense seal timepiece has a much earlier history than would be anticipated, and dates from at least between the fifth and the sixth centuries A.D.

MODERN INCENSE-SEAL TIMEPIECES

It is a matter of considerable interest to discover that the incense-seal timepiece has survived in China to the present time. Examples are little noted even among the modern Chinese, and it is very likely that the timetelling function of these utensils is forgotten and unknown in recent times. The relatively modern examples which have been studied consist merely of a container of metal, usually footed, with five sections: a base, a round or four-sided body, a tray for the ash bed, a perforated grid pattern and a perforated cover. Utensils include a tamper, and a little shovel. Typical examples of modern incense seals are illustrated in figure 11.

The lower compartment apparently served for storage of incense supplies. The tools are stored in the upper compartment, formed by the bottom of the ash bed and the partition. When the timepiece is to be used, the ash bed is covered with a layer of finely sifted wood ash, tamped into place with the tamping utensil. The perforated grid pattern is placed over the ashes, and the pointed end of the little shovel is used to mark out the path, making a V-shaped groove. Powdered incense is then ladled into the grooves by means of the shovel utensil and smoothed

FIG. 11. Modern Chinese incense seals from the
collection of the writer.

into place to the level of the top of the grid pattern, which is then removed. Little bamboo pegs marked with the appropriate hour symbols are placed at regular intervals along the flame track or incense trail. It will be noted that the incense trail is continuous from a starting point. The incense is ignited at the starting point, and the flame track will burn continuously following the intricate lines of the pattern, at an even rate and without flame. Once the incense has begun to burn, the perforated cover is placed over it to protect it from drafts.

FIG. 12. Mdoern incense seal from the collection of the
Science Museum, South Kensington, London.

From more than a dozen of these devices which were collected and studied by this writer, and others which have been brought to his attention, certain basic characteristics have been determined:

The Chinese incense-seals which have survived are invariably made of metal. Of twelve examples studied at first hand, ten were made of Chinese white copper or "Paktong," one was of pewter and another was made of bronze. Only four examples had maker identification or hallmarks. The size of the devices ranged generally between three to four inches square or in diameter, and 4 to 6 inches in height. There is considerable variety in shapes, including square, rectangular, circular, and other forms.

"Paktong" is an alloy of zinc, nickel, and copper

which has been extensively used in China for at least two thousand years before nickel was isolated. The Chinese named the alloy "peh tung" meaning "white copper" because of its appearance, and from this name the Westernized "Paktong" was derived. The metal is known also as "tutenag," and resembles German silver. Like brass and copper, Paktong is cast in sheets by the Chinese, which are then cut into

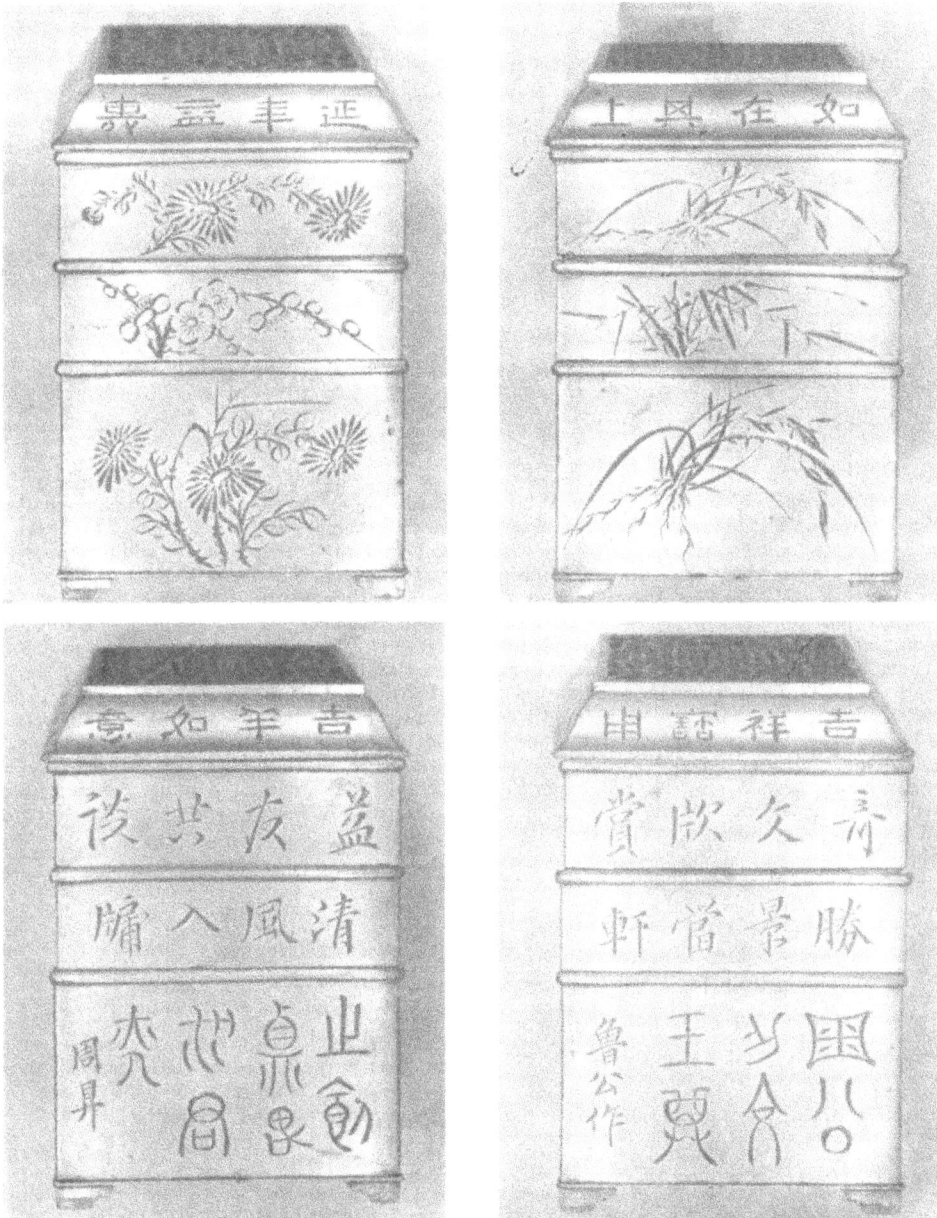

FIG. 13. Modern incense seal from the collection of Dr. William Barclay Stephens.

(Continued on next page)

FIG. 13. (*Continued*)

suitable strips to form the parts of the incense seal and then skillfully soldered together. A raised beading of red copper ordinarily is used along the edges of the major sections for decorative effect.

Occasionally, two incense patterns are furnished with an incense seal. Such is the case with the example shown in figure 12. In such cases, the design of the patterns or grid is different. Presumably each had a special significance and was used accordingly. The incense seal shown in figure 12 is hallmarked on the under side with the characters for "Chao Ch'eng Ming Hsin" which may be translated as "Made by Ming Hsin in the town of Chao."[28]

Another hallmarked example shown in figure 13 is marked on the underside with the symbols for the legend "Ch'ao-yang (district). The shop is in Swatow—'The Genuine Old Shop'—Genuine material tin." This example is made of a material which more closely resembles fine pewter than Paktong. The container is decorated with a beading of brass, and the sides are decorated with engravings of decorative designs and inscriptions.[29]

Another incense seal[30] which appears to be of approximately the same period and origin as the two previous examples is shown in figure 14. It is square in form with three sections in addition to the perforated cover. The device is made of pewter except for the perforated grid of the cover. This is made of brass, presumably so that it will not be affected by heat. The top edges of each section are decorated with thin brass beads.

It is made almost entirely of pewter, including the two small tools. The handles of the pattern tray, and the central part of the upper grid are made of brass. The outer surfaces of the seal are finely engraved with a variety of motifs and inscriptions as illustrated. A free translation of the inscriptions follows:

a) Cleanse the mind and smell [listen to] the wondrous incense;
b) Official rules are impartial;
c) Treasure this incense burner through the generations;
d) The lasting fragrance of the incense resembles spring;
e) The fragrance of the incense goes through my red sleeves as I read the book at night;
f) Use with care this auspicious burner.

Each part of the incense seal bears a hallmark stamped into the underside, which has been translated as "The Chou State arose brilliantly. She is now strong enough to think of launching a war." From

---

[28] From the collection of the Science Museum, South Kensington, London. Illustrated with permission.

[29] From the collection of Dr. William Barclay Stephens of Alameda, California. The inscriptions on each side reading from top to bottom are as follows: (*a*) "Good fortune and use it as treasure"; "Appreciation of extraordinary prose"; "Excellent scenery faces the Hall"; and "Duke Chou makes Vessel for King Wen, made by Duke of lu (son of Duke of Chou)." The line in the next view is (*b*) "Increase years and add longevity"; the third view is (*c*) "Good luck and wishes come true"; "Conversation with helpful friends"; "Clear breeze enters window"; and a garbled inscription from an archaic Chou *t'ing*. The line which appears on the topmost section in the fourth view

is (*d*) "As if the deities were above." The motif of the perforated grid is the symbol for double happiness. Quite probably the sections were assembled so that inscriptions and engraved decorations alternated on each side. This example measures three inches square and four inches in height.

[30] From the collection of the author.

FIG. 14.   Modern incense seal, from the collection of the writer.

the modern incense seal and figure 17 reproduces an example which deserves special mention because it is formed as a cylinder. The grating of the pattern for the incense path is shaped as a "shou" or symbol for long life.

The variety in shapes of the incense-seals provides an interesting study in itself. Decoration of the exterior is ordinarily limited to the minimum, but occasionally the sides are engraved, as in the instance of the seal shown in figure 17, in which the upper sections have designs in archaic characters representing felicitous family wishes, while the mid-section is engraved with symbols for the ancient Chinese "household furniture" and the lower parts are engraved with a continuous landscape.[33]

In a comparison of a dozen and more examples of the Chinese incense-seals of the modern period, one startling similarity is revealed. The pattern used as an incense trail is the same in nine of ten examples! Each of the nine patterns utilize a stylized form of the single character *shou* meaning "long life" or "longevity," or of the two characters *yen nien* signifying "prolonged years." The *shou* character is probably the most popular one among the Chinese people and has been used as a decoration from ancient times. The love that the Chinese people have for presenting and receiving gifts expressing happy augury, which feature the design of one or more of the seal characters for "happiness," is well known. Although the characters for happiness, good luck, wealth and longevity are most frequently encountered, the character for longevity is the one for which there is the greatest preference. This character or *shou* has been formed into more than one hundred variations, some of which have become basic forms in Chinese design and decoration.

The symbol for longevity is commonly encountered also as the design on the central loose button of the Buddhist prayer wheel carved in white jades.

This feature which so many modern incense seals have in common points up the unusual selection of design of the example shown in figure 18. The exterior parts of this seal are made of bronze, while the bottoms of each section are made of brass. It is shaped in the form of a gourd or melon, and the pattern for the incense trail consists of the symbol *mien* meaning "gourd" or "melon" doubled in Siamese fashion. Unlike the other examples noted, the grid in the cover which permits the smoke to be emitted does not have a geometric pattern but is pierced with figures of butterflies and gourds in addition to the characters of an inscription. The inscription is in mirror writing and appears in reverse. The characters perforated in the cover are "*kua tieh mien mien*" or "The melon [gourd] vine is long and extended." This is an echo of the first line of *Mao, No. 237*, one of the poems from the classical anthology

FIG. 15. Modern incense seal in the form of a Ju'i scepter, from the collection of Dr. William Barclay Stephens.

these phrases it seems quite likely that these more ornate examples of the incense seal may have been wedding gifts presented to a newly married couple, to be used on the Buddhist altar in their new homes.

The modern incense seal is not always square or oblong. Figure 15 illustrates an example in Paktong shaped like a Ju'i scepter, a symbol which derives its shape from the famous sacred fungus called *ling-chih* (*Polyporus lucidus*) which is one of the Taoist emblems of longevity.[31] A second example in the form of a Ju'i scepter, also made of Paktong, is illustrated in figure 16. This is a particularly fine seal, with the perforated cover decorated with an archaic Chinese inscription. It is believed to date from the Ming period, and is particularly fine in execution throughout.[32]

The cylindrical form is occasionally encountered in

[31] From the collection of Dr. William Barclay Stephens.
[32] Owned by Mr. Ed. S. Jones of Los Angeles, California.

[33] From the collection of Mr. Ed. S. Jones, Los Angeles.

Shih Ching (The Book of Songs).[34]   The poem is an ancestral hymn of the Chou Dynasty, wherein the line occurs as *"Mien mien kua tieh."*   The traditional interpretation is that just as the melon or gourd close

---

[34] The *Shih Ching* was probably the most important of the Chinese literary classics, and consisted of 311 poems, most of which were produced during the Middle Chou period (900 B.C.–

to the base of the vine is usually very small and that melon at the end of the vine is ordinarily very large, so the founders of the Chou Dynasty had very obscure origins but eventually produced a great country.

The hallmark on the underside, which is also in mirror writing, reads *"Tsung cheng chai tsao"* which may be translated as "Made by the Studio of Ancestral Rectitude" or by an individual named Tsung-cheng chai.

The symbol of the gourd or melon signifies magic and assumes special meaning in Chinese decoration. It is the symbol of Li Tieh-kuai, one of the legendary Taoist Eight Immortals (*Pa Hsien*) who devoted himself to the practice of magic and necromancy.

This incense seal depicted in figure 18 is probably

---

650 B.C.).   Some are sacrificial songs which were probably sung originally to music during ritual dances, while other pieces were of political and/or legendary topics.   Most of them were purely lyrical and dealt with love.   Although commentators believe that water clocks or clepsydrae were mentioned in these odes, modern writers consider the possibility doubtful.   Arthur Waley interprets the line from *Mao, No. 237* as a reflection of the widespread belief among many Asian and African peoples that man had his origin in melon seeds, but this is not the translation traditionally accepted in China.

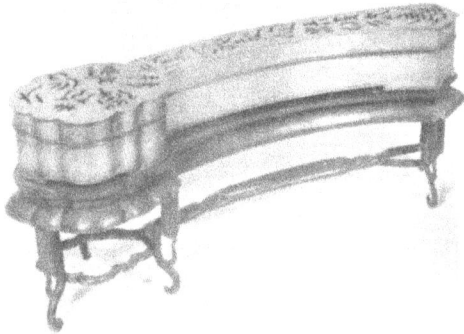

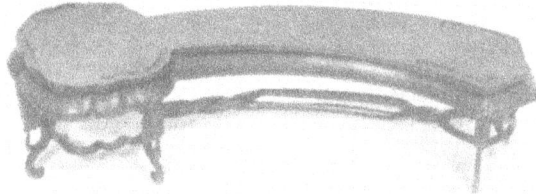

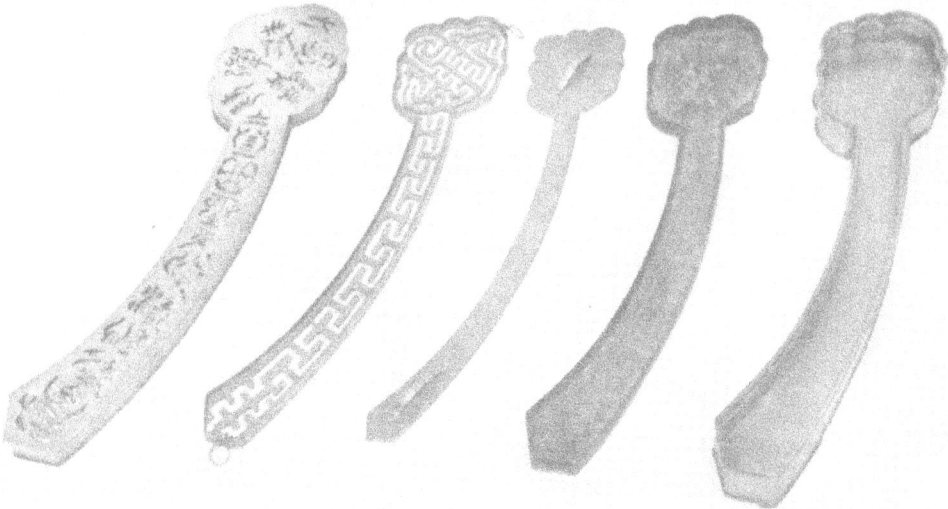

FIG. 16.   Modern incense seal formed like a Ju'i scepter with archaic inscription, from the collection of Mr. Ed S. Jones.

the oldest of the examples studied, and appears to
date from the first half of the seventeenth century.
It is believed that these incense seals served a function
in the Taoist and Buddhist temples to measure the
time periods for striking the great temple bell which
notified the villagers that it was again time for prayer.

FIG. 18.   Bronze incense seal in shape of a melon or gourd, from
the collection of the writer.

It seems to be a logical conclusion that the introduction of European timepieces into China by missionaries
in the seventeenth century, and the considerable
traffic in clocks and watches from Europe to the
Orient which developed in the ensuing periods, made
the incense seal impractical as a timepiece and that it
continued merely as a vestige in ceremonial form.

It is difficult to determine how or when the various
examples of incense seals noted in the foregoing were
exported from China.   It is a natural conclusion that
they were collected and purchased by soldiers and
tourists as curios.   When examples have appeared in
the shops of curio and antique dealers, they have been
"authenticated" variously as opium stoves, hand-
warmers, and charcoal carriers but never as timekeepers.

MISSIONARY ACCOUNTS OF CHINESE INCENSE TIMEKEEPERS

The employment of incense for time measurement
in China was verified in the writings of European
missionaries and emissaries who visited China in the
sixteenth century and later.

The first of these references occurred in the *Memoirs*
of Father Matteo Ricci (1552–1610), the founder of
the Jesuit missions in China.[35]   In a journal entry
dating probably after 1601 while he was still residing
at Peking, Ricci wrote that

As for their clocks, there are some which use water, and
others [which use] of certain perfumed fibres
made all of the same size; besides this they make others
with wheels which are moved by hand—but all of them
are very imperfect. . . .

A relatively free rendering of Ricci's journals was
made by Padre Nicolas Trigault in his account of the
Jesuit missions in China, which was published
simultaneously in Vienna and Augsburg in 1615.[36]
With reference to Ricci's entry on timekeepers quoted
in the foregoing, Trigault rendered it as follows:

Horis metiendis vix habent instrumenta, quae habent,
vel aqua vel igne mensurantur.   Aquea sunt velut in-
gentes clepsydrae; Ignea ex odorifero cinere confecta,
tormentorum nostrorum fomites imitantur. . . .

This passage in Latin from Trigault's work has
appeared in translation in several published works,
each with a somewhat different interpretation.   In
the translation of Trigault's work published in 1953
by Rev. Louis J. Gallagher, S.J.,[37] the passage appears
as follows:

This land possesses few instruments for measuring time
and in those instruments which they have, it is measured
either by water or by fire.   The instruments run by
water are fashioned like huge waterpots.   In those which

[35] Pasquale D'Elia, *Fonte Ricciane* (Rome, 1942) 1: pp. 33.
[36] P. Nicolas Trigault, *De Christiane Expeditione Apud Sinas*
(Augsburg, 1615), p. 22.
[37] Rev. Louis J. Gallagher, S.J., *China in the Sixteenth Century;
the Journals of Matteo Ricci* (New York, 1953).

are operated by fire, time is measured by an odoriferous ash, somewhat in imitation of our reversible grates through which ashes are filtered. A few instruments are made with wheels and are operated by a kind of bucket wheel in which sand is employed instead of water, but all of which fall short of the perfection of our instruments, are subject to many errors, and are inaccurate in the measurement of time. . . .

In their recent work entitled *Heavenly Clockwork*,[38] Drs. Needham, Ling, and Price proposed another version, as follows:

They (the Chinese) have very few instruments for measuring time, and those which they do have measure it either by water or by fire. Those which use water are like large clepsydrae, and those which use fire are made of a certain odoriferous ash very like that tinder which is used for torture among us. . . .

This second version was based on the premise that Trigault was referring to moxa or medical cauterization, inasmuch as the practice of using incense ash for this purpose was characteristically Chinese.

In order to obtain clarification of this significant passage, the present writer corresponded with Rev. Gallagher with most satisfactory and definitive results. Father Gallagher suggested that the difficulty in the passage derives from the words "tormentorum" *(tormentum)* and "fomites" *(fomes-fomitis)*. The basic meaning of the Latin word *tormentum* is any machine or device for twisting or turning any object. Inasmuch as the common appliance for turning or filtering ashes is a grate, he gave this application to the word in his published translation.

When consulted by Father Gallagher, Father Joseph Sebes, S.J., Professor of Chinese History in the graduate courses at Georgetown University, proposed a translation which is probably even more definitive than the others. He has taken the word *fomites* of Trigault's text to mean "tinder, touchwood used for the purpose of igniting gunpowder." He took the words *tormentorum nostrorum* of the same passage to mean "our guns" and rendered the passage as

. . . In those [instruments] which are operated by fire, time is measured by an odoriferous ash, like unto the tinder-sticks or touchwood of our guns.

As his authority for his translation, Father Sebes mentions that Fr. Johann Adam Schall Bell, a Jesuit missionary to China in 1665 used the term *"tormenta bellica"* to mean "machines of war" or "guns." [39]

Although the passage may have referred to the incense-seal timepiece, it is far more likely that Father Ricci was describing the incense sticks divided with markings of time periods.

Undoubtedly the simplest form of incense timepiece was the match-cord, consisting of a length of

specially prepared cord of punk, knotted at measured intervals which indicate the passage of time by the progress of the burning. Punk burned slowly and evenly, and the knotted cord was used among the poorest classes in China as a time alarm. A length of punk or match-cord, knotted for the period of time desired, was placed between the bare toes and ignited by a Chinese before he fell asleep. When the cord had burned down to the skin, it would awaken him, and undoubtedly he would unwittingly awaken the rest of the household.

The use of the match-cord, perhaps in a more elaborate form, was noted by Nieuhoff in his report of the embassy of Peter de Goyer and Jacob de Keyser of the East India Company of the United Provinces to the Grand Tartar Cham, Emperor of China, in 1665.[40] Nieuhoff related that

The Chinese, although most ingenious and subtle, have no accurate instruments to indicate time, and those which they have are as imperfect as can be made. Those which indicate the hours by means of water resemble in some ways our pounce boxes, and those which are by means of fire resemble our wicks (or fuses). There are some who attempt to make sundials, but with so little success that it is a pity. . . .

A refinement of the match-cord was described several years later, in 1668, in a work by Padre Gabriel de Magalhaens, S.J.[41] Padre Magalhaens (1611–1677) was one of the most outstanding of the Jesuit missionaries who arrived in China in 1640 and he spent the last thirty-seven years of his life working among the Chinese. He constructed several clocks for the Chinese emperor and he naturally became interested in the indigenous forms of timetelling. In his published work he stated:

The Chinese have found, for the regulation and measurement of the parts of the night, an invention worthy of the marvelous industry of this nation. From a certain wood, which they grated and pounded to reduce it into a powder and form from it a paste, with which they form into cords and sticks of various shapes. Some are made from more precious woods, such as sandalwood, *bois d'Aigle* and other odorous woods, with a length of about one finger or thereabouts, which the rich and literate personages caused to be burned in their chambers. There are other, more ordinary, forms of one, two and three cubits of length, and also of one, two or three ells, and a little more or less larger than a goose quill, which they burn in front of their pagodas or idols. They use them also as a match (wick or fuse) to bring fire from one place to another.

They make cords of powdered wood of equal size, uniform thickness, by passing them through a draw-plate or a trough specially made for the purpose. Then they coil them into a round form by beginning at the center and forming a spiral or conical figure which enlarges with each turn, up to one, two or three palms in diameter, and even more, and which will have a duration of one, two or three days or even more, in proportion to the size,

---

[38] Joseph Needham, Wang Ling and Derek J. Price, *Heavenly Clockwork* (Cambridge, 1959), p. 155.

[39] The translation made by Father Sebes was approved by Father Neil Twombley, S.J., Latinist in the School of Languages and Linguistics at Georgetown University.

[40] John Nieuhoff, *An Embassy from the East India Company* . . . (London, 1673), p. 159.

[41] P. Gabriel Magalhaens, *Nouvelle Relation De la Chine* (London, 1668), pp. 153–154.

FIG. 19.   Incense spiral in foreground.   Chinese interior from Favrier.

because one sees them in the temples which last ten, twenty or thirty days. These machines or matches resemble a fisher's bow net, or a cord wound around a cone. They are suspended from the center and they are lighted at the bottom end, from which the smoke issued slowly and faintly, following all the turns which has been given to this coil of powdered wood, on which there are ordinarily five marks to distinguish the five parts of the evening or night. This method of measuring time is so accurate and certain that no one has ever noted a considerable error. The literate, the travellers, and all those who wish to arise at a precise hour for some affair, suspend at the mark which they wish to arise at, a small weight which,

FIG. 20.   Table model of incense spiral, from Planchon.

when the fire has arrived at this spot, invariably falls into a basin of brass which has been placed below it, and which awakens the sleeper by the noise which it makes in falling. This invention takes the place of our alarm clocks, with the difference that they are very simple and extremely inexpensive that one of these machines that can last for a period of twenty-four hours, costs no more than three deniers, and that the clocks are composed of a quantity of wheels and other pieces and are consequently very expensive that they cannot be employed except by the rich personages.

Thus Father Magalhaens described not only the incense stick but the incense coil, both of which are produced of hardened paste incense, and both of which have played important roles in the story of timetelling with incense in the Orient.

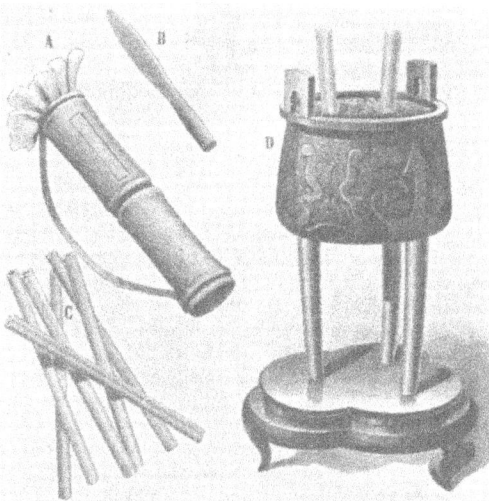

FIG. 21.   Timetelling with incense sticks, from Planchon.

The description of the incense coil or spiral requires no amplification. These coils have been noted by travelers in the temples and shrines of southern China and French-Indo-China even in modern times in great profusion, and they are commonly available for purchase in these areas.

Figure 19 reproduces a drawing in the work of Mons. A. Favrier of a Chinese interior in which only a clock of European type is featured, but an incense spiral may be distinguished in the foreground on the floor.

The incense spiral is ordinarily hung from a rafter or from the ceiling and is so shown in a photograph of the interior of a temple in French Indo-China in the work entitled A Dragon Apparent by Norman Lewis.[42]

The incense spiral could be used on a table also,

[42] Norman Lewis, A Dragon Apparent: Travels in Indo China (London, 1951), p. 177.

FIG. 22.   Incense sticks in container, from Chambers.

and it is shown in this form in an illustration in Planchon's work, which is reproduced in figure 20.[43]

In his recent work entitled *Das Sanduhr Buch*,[44] Ernst Jünger noted that incense was used in spiral form when a longer period of burning was necessary than was possible with the common incense stick. These spirals served to measure the "night watches" primarily, and therefore the length of the spirals varied with the months during which they were employed. One circle of a spiral lasted for the period of a single night. Each of the circles was divided evenly into intervals and marked to designate the reliefs of the five night-watches.

In addition to the use of incense in the spiral form for time measurement, incense was employed in stick form for the same purpose from ancient times. The burning time for the sticks of hardened incense paste was noted and the incense sticks marked off along their lengths with time indicators. In the same manner that these incense sticks were also used as votive offerings before shrines and upon the Buddhist altars, the sticks were inserted vertically into a burner (*lu*) half filled with wood ashes. The passage of time could be easily noted as the stick was consumed. An engraving of the incense stick in a burner is re-

produced in figure 21 from Planchon's work. This engraving is based on an earlier illustration which appeared in a work by Chambers published in Paris in 1757,[45] and which is shown in figure 22.

In a report on the Chinese timekeeping methods[46] which he submitted to the Smithsonian Institution in 1851, a medical missionary in China named Dr. D. J. Magowan stated:

> Time is often kept with tolerable accuracy in shops and temples by burning incense sticks made of sawdust, carefully, but slightly, mixed with glue, and evenly rolled into cylinders two feet long, and divided off into hours. When lighted, they gradually consume away without flame, burning up in half a day.

Alice Morse Earle[47] reported that the guardian of the "copper-jar-dropper," the famous water clock which stood in the watch-tower at Canton, sold "time-sticks" or incense sticks marked for timetelling as late as 1899. She went on to describe these time sticks as being "made of sawdust (usually of a certain wood), a slight mixture of glue, rolled into even cylinders two feet long, and divided into hours. They consumed without flame, and burned up in half a day." The similarity in description with the words of Dr. Magowan is to be noted.

Earle quoted an old Chinese poem known as a "short stop" in which a waiting wife complains of the leaden foot of Time. The water-clock and the incense sticks are described as timekeepers in the poem, which is translated as follows:

> It seems that the Clepsydra
>   Has been filled up with the Sea
> To make the long, long nights appear
>   An endless time to me.
>
> The incense-stick is burned to ash
>   The water-clock is stilled
> The midnight breeze blows sharply by
>   And all around is chilled.

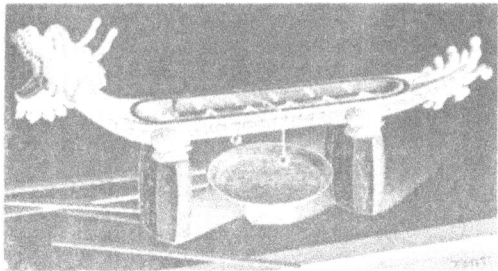

FIG. 23.   Dragon vessel timepiece, from Planchon.

[43] Mathieu Planchon, *L'Horloge: Son Histoire Retrospective, Pittoresque et Artisique* (Paris, 1899), p. 255.

[44] Ernst Jünger, *Das Sanduhr Buch* (Frankfort, 1959), pp. 49–52.

[45] Sir William Chambers, *Traité des Edifices, Meubles et Habits de la Chine* (Paris, 1757).

[46] Dr. D. J. Magowan, "On Chinese Horology," Report of the (U. S.) Commissioner of Patents for the Year 1851 (Washington, 1852); "Modes of Keeping Time Known Among the Chinese," *Chinese Repository* (July, 1851), p. 611.

[47] Alice Morse Earle, *Sundials and Roses of Yesterday* (New York, 1902), p. 54.

FIG. 24.   Dragon vessel of lacquered wood, owned by
Miss Helen C. Hagar.

THE DRAGON INCENSE VESSEL

One of the most interesting forms in which incense was used for timekeeping by the Chinese was in the employment of the so-called "dragon boat" in which one type of incense stick was burned. One example of such a vessel from the collection of the Musée de la Marine at the Louvre in Paris was illustrated and described in Planchon[48] and reproduced here in figure 23. A similar vessel owned by Miss Helen C. Hagar of Salem, Mass., is shown in figure 24. This dragon boat measures about 27 inches in length with a beautifully carved dragon's head at one end which displays a red tongue. The body is of black lacquer on wood with gilded scales and it terminates with a tail painted flame red. The four little legs conceal tiny brass wheels which permit the vessel to be rolled about on the surface of a table. A pewter liner inside the vessel has nine wire seats shaped in V form which serve as a rack for a lighted incense stick.

According to Planchon, the vessel served as the holder for an incense stick used to measure time during the night. Two tiny metal weights were suspended between a fine string across the body at the position desired so that the string would burn at the designated hour when the consumption of the incense stick by burning reached that point. As the burning incense caused the string to part, the little weights would fall into a metal dish placed beneath the vessel and the sleeper would be awakened by the sound. In Planchon's drawing the string is stretched at too great a distance from the burning incense to make this arrangement successful, but this is undoubtedly the artist's oversight, and the discrepancy is easily remedied.

The "dragon boat" is described and illustrated also in Robert Guitton's *Quand Sonne L'Heure* where it is called "la Jonque a repitition" or "the repeater junk." It is described as a Chinese implement, and the first example of a morning alarm. One of these devices which was seen by Guitton was made of bronze, measured about 80 cm. in length with twenty straight rods along the open center upon which an ignited incense stick was laid. Then from edge to edge thin strings were laid at equal intervals in contact with the incense stick at spaced intervals

along its body. To the ends of each of these twenty strings were attached tiny bronze balls which hung below the bow of the vessel. As the burning of the incense stick progressed, the strings would part one by one and the little balls would drop into a bronze platter over which the bronze junk was suspended.

Later, when timepieces of other forms became much more common in China, these timetelling junks were apparently utilized for other purposes, for in his book entitled *The Canton Chinese, Or The Americans' Sojourn In The Celestial Empire*, which appeared in 1844, the author, Osmond Tiffany, Jr., wrote that the owners and counting-house clerks of the nineteenth-century *hongs* of Canton partook of dinner together after the end of a business day. When the wine was being served, the "dragon boat," laden with unlighted joss-sticks was sent up and down the table, and the joss-sticks used to light the fine Manila cheroots which ended the dinner.

The "dragon boat" or incense junk, as well as the incense sticks, were included in the comprehensive study of timetelling in China by Professor Alfred Chapuis entitled *La Montre Chinois*,[49] in an account which appears to be a summation from Planchon and other writers:

As a measure of time, the Chinese rather used fire clocks (*horloges a feu*) made by means of the same mixture causing combustion of which such great use is made in China in the form of odoriferous sticks which are con-

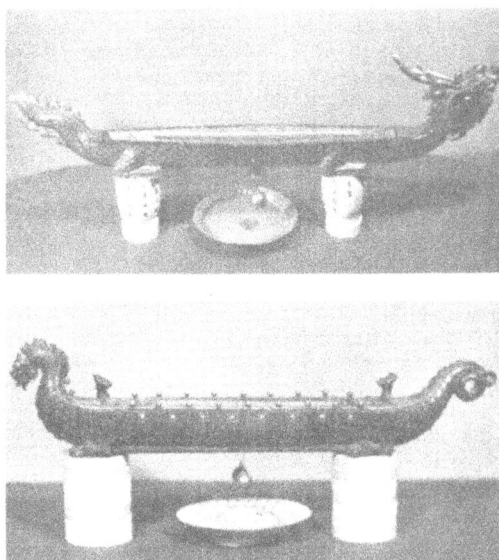

FIG. 25.   Two dragon vessel timepieces, one of wood and
one of bronze, from the collection of Dr. S. P. Lehv.

---

[48] Planchon, *op. cit.*, pp. 257–259.

[49] Alfred Chapuis, *Le Montre Chinois* (Neuchatel, 1918), pp. 16–17.

tinuously burned before the tablets of ancestral worship or their idols. This mixture, with a base of clay, is composed of the sawdust of the wood of various species of aromatic trees from Tibet, to which are added musk and gold dust. These sticks, which consume themselves slowly without ever igniting, are occasionally marked with gradations to serve as time measurers; their burning can last for several days and they indicate the time with sufficient accuracy. These fire blocks assume several forms and can also serve as alarms (i.e., the dragon vessel).

As an interesting postscript to the ancient use of incense sticks in China for time measurement is the account of its use even to the present time. Rudolph P. Hommel, in his important study of Chinese tools and implements entitled *China At Work*,[50] stated that coal miners, who work underground continuously for consecutive periods of about three hours each, carry what they call a "Timepiece" in order to tell the time and measure their period of work. This is simply an incense stick which glows for three hours, without flame.

The inexpensive method of burning incense-sticks to tell time continued in common practice in the early years of the Chinese Republic. An expression that is often encountered in Chinese books is *"i-chu hsiang te shih-hou"* or "the time of burning an incense stick," as exemplified in *Chung-shan ta-tzu-tien i-tzu ch'ang-pien.*

An interesting application of the incense stick for time measurement was brought to this writer's attention by Dr. M. K. Starr, Curator of Asiatic Archeology and Technology at the Chicago Natural History Museum. Dr. Starr noted that it was a practice in former times—and perhaps even today—for older men, scholars, to sit about together and compose poetry upon a mutually selected theme. A given period of time was designated for the composition of the poetry. The time was measured by means of an incense stick to which a small bell was attached by a string at the designated point. When the incense had burned down to the string, the string parted and the bell fell, indicating to the would-be poets that the time was up.

INCENSE TIMEKEEPERS OF THE JAPANESE

It is not surprising that the methods of employing fire and incense for time measurement in Japan closely paralleled those evolved by the Chinese. It is a natural assumption that the Japanese borrowed from China, based on the general premise that so great a part of Japanese culture was acquired from China between the sixth and the ninth centuries. Quite often the Japanese developed their borrowed arts and skills to a much greater degree than had been achieved by the originators. An example of this is the welding of hard and soft steel for swords, which

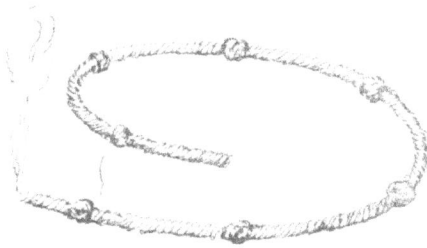

FIG. 26.   Timetelling with a knotted match-cord.

was known in China as early as the third century. It did not make its appearance in Japan until the eighth century, but during the thousand years that followed, the Japanese perfected this type of metalworking to a degree that the Chinese never achieved.

THE MATCH-CORD

According to the *Tokei Hattatsu-Shi* (Development of Clocks) by Hyoe Takabayashi, punk or match-cord was used by the Japanese in the same manner that it had been employed in China for timetelling purposes, and presumably it was borrowed from the Chinese (fig. 26).

THE CANDLE TIMEPIECE

Candles were employed in Japan as well as in China for time measurement, and the use of the so-called graduated candle for this purpose was described in the well-known Chinese-Japanese encyclopedic dictionary, the *Dai Kanwa Jiten*.[51] Hideo Kunitomo listed it as one of the Japanese *Hidokei* ("fire clocks"). He indicated that inasmuch as the speed of the candle's consumption by fire varied according to the materials of which the candle was composed, it was not successful as a standardized timepiece.

In Japan as in China, the graduated candle was used, segmented along its length by horizontal markings to indicate the time periods. According to Takabayashi, the scaling of the candle was grooved out and the resulting channels filled with a white, bone-like powder. The same source stated that such candles were used until A.D. 901.

JAPANESE INCENSE-STICKS

The Japanese adopted the use of the incense stick or joss-stick for timetelling as a matter of course. Known in Japan as "Chinese matches," the most common type were made of powdered sandalwood reduced to a paste and hardened. At first the period of time required for a stick's consumption was observed and used as a time interval. The device was later refined by marking the sticks at measured intervals with time designations as in China.

[50] Rudolph P. Hommel, *China at Work* (Doylestown, Pa., 1937), p. 4.

[51] *Dai Kanwa Jiten* **2**: p. 264b.

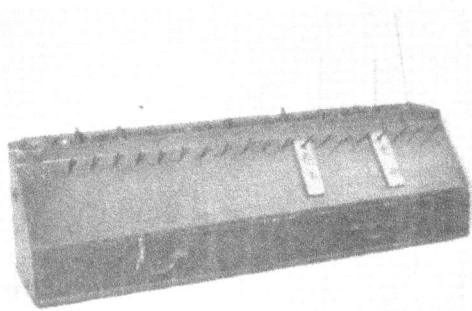

Fig. 27a.   Geisha timepiece, from the collection of the
National Science Museum in Tokyo.

THE GEISHA TIMEPIECE

Until quite recently incense sticks were utilized in a
special type of timepiece used exclusively in geisha
houses (*Okiya*), to compute the cost of the entertainment.   Incense sticks which had a burning duration of
one half-hour were used, and kept in a special box
(fig. 27a) with a number of openings in the top
designed for the insertion of incense sticks in bamboo
holders.[52]   The interior of the box was in the form of
a drawer for the storage of incense supplies.   Each
opening was assigned to one of the girls of the establishment and it was customary to mark each with a

tag bearing the girl's name.   As each girl became
employed during the evening to provide entertainment for a caller, an incense stick was attached into
a bamboo holder, lighted, and dropped into the hole
in the box opposite her name.   As the evening
progressed, and the sticks had to be replenished, a
soroban (fig. 27b) or abacus was used to compute the
number of incense sticks consumed for each visitor.[53]
According to Takabayashi, geisha entertainment as
late as 1924 was being computed and charged on the
basis of incense sticks, each of which burned for a
half-hour period.

Ernst Jünger amplified the description of the geisha
timepiece in a recent work, with details he had ob-

Fig. 27c.   Geisha timepiece, in the collection of the writer.

tained from Giko Takabayashi.   Jünger described
the timepiece as

. . . a wooden box in which nine holes are drilled: one
for each of the girls living in the house.   Each time one
of the girls retired with a guest, the host (or senior in
charge of the geisha house) would light an incense stick
and set it up for her. . . . The duration of burning corresponded to a definite amount.   A flower girl could say
"Yesterday I earned six sticks!"   Today such boxes are
no longer in use.   But the Japanese word for the sticks,
"*senko*," still retains its meaning, so that even today one
asks what is the price in this or that district for a "flower
girl incense stick."   To such elegance of the blossom
language, we Western barbarians would never penetrate.

A geisha timepiece[54] is shown in figure 27c.   It is
made of *sugi* or cryptomeria wood and measures $8\frac{1}{2}$
inches long by $4\frac{3}{4}$ inches wide by $1\frac{3}{4}$ inches high.
Twenty holes are drilled into the top in two rows of
ten in each row, for the insertion of incense stick
holders.   The drawer of which the interior consists is
divided for the storage of incense sticks and holders.

The geisha timepiece probably originated with the
establishment of the first geisha houses.   The geisha

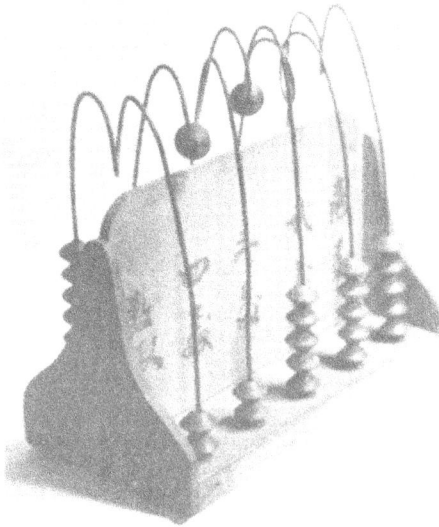

Fig. 27b.   Soroban for computing number of incense sticks, in the
collection of the National Science Museum in Tokyo.

[52] From the collection of the National Science Museum in
Tokyo.   Illustrated by courtesy of Dr. Teiichi Asahina.

[53] Soroban from a geisha house, in the collection of the National
Science Museum in Tokyo.   Illustrated by courtesy of Dr.
Asahina.

[54] From the collection of the writer.   Obtained in Tokyo.

houses came into being at the beginning of the Edo period in the early seventeenth century. The use of these timepieces was probably discontinued late in the nineteenth century, when the westernization of the islands began.

"INCENSE SEAL" TIMEPIECES IN JAPAN

The Japanese adopted the Chinese "incense seal" timepiece at an early period, and made only several slight modifications. First of all, they used boxes or containers made only of wood instead of metal. The boxes were considerably larger than those employed among the Chinese. In every instance, the wood used appears to have been the same, a strongly grained hardwood that resembles pine in color and structure, and which is probably a species of cryptomeria. The exteriors of the containers were lacquered in clear, red or black lacquers or in combinations of two or more.

The earliest forms of the Japanese incense clocks were extremely primitive. They consisted of a solid cube of hardwood, measuring about twelve to fourteen inches square on each facing. On the top surface was carved a continuous pattern that occupied the entire surface. This pattern consisted of a channel square in form about $\frac{3}{16}$ of an inch in width and depth. Powdered incense was filtered into the pattern and ignited at a starting point. The incense burned its way along the path to its conclusion.

CAPTAIN GOLOWNIN'S ACCOUNT

An interesting use of incense was reported by Captain Golownin[55] in an account of his captivity in Japan early in the nineteenth century. He stated:

To measure time, they (the Japanese) employ a small beam (block) of wood, the upper part of which is covered with glue, and whitewashed; a narrow groove is made in the glue and filled with a vegetable powder, which burns very slowly; on each side of this groove, at certain distances, there are holes formed for the purpose of nails being put into them. By these holes, the length of the day and night hours is determined for the space of six months, from the spring to the winter equinox. During the other six months, the rule is inverted, the day becoming night hours, and the night day hours. The Japanese ascertain the length of a day hour, and mark it off with nails; they then fill the groove with powder, set light to it at noon, and thus measure their time. The beam (block) is kept in a box, which is laid in a dry place; but the changes of weather have, notwithstanding, a great influence on this kind of timekeeper.

Galownin's device may be identified with a primitive form of the incense seal, for the "vegetable powder" was undoubtedly incense.

THE JO-KO-BAN, OR "PERMANENT INCENSE BOARD"

According to Dr. Asahina and Professor Yabuuti,[56] there are two basic types of incense timepieces in Japan, of which the first is the *Jo-ko-ban* or "permanent incense board." This was used in many Buddhist temples, designed for and understood by the initiates of the priesthood. Its purpose was primarily the continuation of the temple's incense fire over a relatively long period, such as the duration of a night or longer. Basically, it was not intended as a device for time measurement. Examples may still be encountered in Buddhist temples in Japan even at present, but they are no longer used for their original purpose. In one of the temples at Kobe one of these devices is carefully preserved, and is known as a *Jo-ko-ro* or "permanent incense furnace."

In the *Jo-ko-ban* the incense pattern is formed in a zigzag line over a leveled bed of finely sifted ashes. The incense trail is ignited at one end and burns for a considerable period of time without requiring attention. Time plates made of bamboo or other materials are inserted at regular intervals along the trail to denote the completion of the hours, as the incense burns. Although it was not originally intended as a timekeeper, the device was gradually appropriated for the purpose of measuring the intervals at which the temple bell was to be rung as a signal for prayer, thus evolving into a timekeeper. Since the amendment of the Japanese calendar in the sixth year of the Meiji period (1873), when the old time system and lunar calendar were abandoned, the time plates became useless and were eventually discarded.

THE JI-KO-BAN OR "TIME MEASURING INCENSE BOARD"

According to Asahina and Yabuuti, the second type of Japanese incense clock was called the *Ji-ko-ban* or "time measuring incense board."

It is of interest to note at this point that the Japanese characters for *ko-ban* which means literally "incense basin" are identical with the Chinese characters for *hsiang-p'an* which has the same meaning in that language.

A *Ji-ko-ban* is used in the one instance in which the incense timepiece continues to play an important role in Japan, namely, during the *Mizu-tori* or spring festival celebrated each year at Nara.

THE MIZU-TORI OR SPRING FESTIVAL

The *Mizu-tori*, which means literally "the water-drawing ceremony," is also known as the *Tai-matsu Shiki* or "pine torch ceremony." The festival begins each year on March 1 to herald the official arrival of spring and lasts for two weeks until March 14.

---

[55] Capt. Golownin, *Memoirs of A Captivity in Japan During the Years 1811, 1812 and 1813* (London, 1824).

[56] Personal communications with Dr. Asahina and Professor Yabuuti.

FIG. 28.   Jikoban used in the Nigatsudô at Nara.

Pilgrims travel from all over Japan to visit the Nigat-sudô temple at Nara during this period.[57]

The festival takes place each year during the two weeks beginning March 1 because it is during this time that water flows from a sacred well in the Nigatsudô, an event which is the basis of the cere-mony. It is believed that an intermittent natural spring lying far back in the hills behind the temple overflows at this particular time because of the melting of the snows in the mountains. The added burden of the melting snows causes the spring to overflow, activating the temple well. This annual phenomenon is given great importance by the Japanese and the water-drawing ceremony has been observed as a religious event for twelve hundred years. The water is collected as the well starts to flow, and dispensed

FIG. 29.   Enlargement of fig. 28, showing indicators for time intervals.

[57] William Hugh Erskine, *Japanese Festival and Calendar Lore* (Tokyo, 1933).

among the pilgrims as miraculous water during the ensuing two weeks. It was estimated one year that seven and a half wagon loads of water were taken away by pilgrims in two and four ounce bottles.

The sacred well is situated in Nigatsudô temple. The name *Nigatsu* means "February" and "-dô" means "Hall." The temple received this name be-cause originally the ceremony took place during February of the lunar calendar (now equivalent to the first two weeks of March). The Nigatsudô forms part of the Tôdaiji Temple at Nara, which was built by the Emperor Shomun (A.D. 724–748).

The *Mizu-tori* was initiated for the first time in A.D. 760 under the Emperor Junnin, and it is probably from that time that an incense timepiece was used in the spring rites, although there is no documentary confirmation. The *Jikoban* which is presently in use is believed to date from the second half of the seven-teenth century. The Nigatsudô was destroyed by fire and rebuilt anew in the Kanbun era and the present timepiece dates from that reconstruction.

Preparations for the *Mizu-tori* begin on February 20 each year. Beginning on the first day of March, the public ceremonies are held continuously day and night until the morning of March 15. The schedule for the festival is based on a time system called *Rokuji* (literally *Roku* meaning 6 and *ji* meaning "times"), by means of which each day is divided into six intervals: *Nittyu, Nichibotsu, Shoya, Hanya, Koya* and *Shincho*. These time intervals are measured in the Nigatsudo by means of the *Jikoban* or incense timepiece.

The *Jikoban* used in the ceremonies is placed in the southeastern side of the Suyadan in the inner room of the temple.

It consists of two boxes placed one over the other. The lower box is 3 *Zun* 4 *Bu* in height, and is used for the storage of incense supplies and such tools as the ash leveler and the pattern tray.

The upper box is square and measures 1 *Shiaku* 2 *Sun* (1 *Shiaku* equals 10 *sun*) on each side, and it is 3 *Zun* 4 *Bu* high. The bottom board is 5 *Bu* in thick-ness, and the bed of ashes is specified to be 1 *Sun* 5 *Bu* in depth.

Figure 28 illustrates the *Jikoban* actually in use in the Nigatsudô. Figures 29 and 30 are enlargements showing the incense trail in close-up detail.

It will be noted that in this instance the pattern for the incense trail is formed approximately in the shape of the letter *W*. This *W*-shaped channel is perforated in the bottom of the pattern tray with which the incense trail is formed. The under side of the tray is carved with eleven parallel gutters which are 3 *Bu* wide and occur at intervals of 7 *Bu*. When the pattern grid is pressed into the leveled bed of sifted ashes, eleven ridges are formed by means of the gutters. These ridges remain on the surface of the ashes and interrupt the incense trail at intervals of

FIG. 30a.   Incense pattern tray, shown from upper side.

FIG. 30b.   Incense pattern tray, showing the bottom side.
(Photos of Jikoban at Nara courtesy of Dr. Asahina.)

1 *Sun*.   The little bamboo markers on peg-like stands to denote the time intervals are clearly visible in the illustrations.

THE KO-IN-ZA IN THE SHŌSŌIN

In the Shōsōin, the famous Imperial Treasury near Nara, there are two *kō-in-za*, preserved in the south warehouse, where they are zealously guarded with the other treasures of the ancient emperors.

The *kō-in-za* means literally "a seat or stand for a censor or fragrant board" or "a seat for an incense seal." The example shown in figure 31 measures 56.0 cm. in diameter and 17.0 cm. in total height. The second example, is 55.6 cm. in diameter and 18.5 cm. high. The *kō-in-za* itself is a round stone stand, as shown in Figure 32, which is placed on a wooden seat or *Gan-za*, shown in figure 33. The *Gan-za* consists of four tiers of plate-like leaves carved of wood to resemble lotus petals. The plates are gilded and painted with patterns and figures on each surface. The paintings represent specific things, such as "Toka" or Chinese flowers, "Ho-so-ka" or flowers of treasure, "Hana-kui-dori" meaning literally "birds feeding on flowers," "Ganju-dori" or birds of honor, "hoh-oh" or Chinese phoenix, "Osidori" or mandarin duck, and "Shishi" or lion.   Each tier includes eight petals.

FIG. 31.   *Kō-in-za* in the Shōsōin at Nara.

FIG. 32. *Kō-in-za* shown from side, not mounted in *Ganza*.

The seat itself is painted brownish green and is inscribed in ink *"kō-in-za"* on the underside. Another name which has been given to it is *Urushi-kinpaku-eban* meaning "a figured plate lacquered and painted gold."

Authorities at the Shōsōin date the *kō-in-za* in the eighth century A.D. It is believed to be of Chinese manufacture, although there is a slight possibility that it is a Japanese copy of Chinese handicraft. The pedestals were originally made and used to contain "incense seals," consisting of Chinese or Sanskrit characters formed from powdered incense over a bed of sifted ashes to indicate the measured passage of time. A record of the *kō-in-za* is to be found in the *"Amida-kohkaryo-shizai-cho"* or inventory of property compiled by Bun-zoh, a Buddhist priest in Japan on the thirtieth of August in the first year of Kei-un. This document is now preserved in the

FIG. 34. *Kō-in-za* in place on *Ganza* seen from above.

FIG. 33. The *Ganza* without the *kō-in-za* viewed from above,

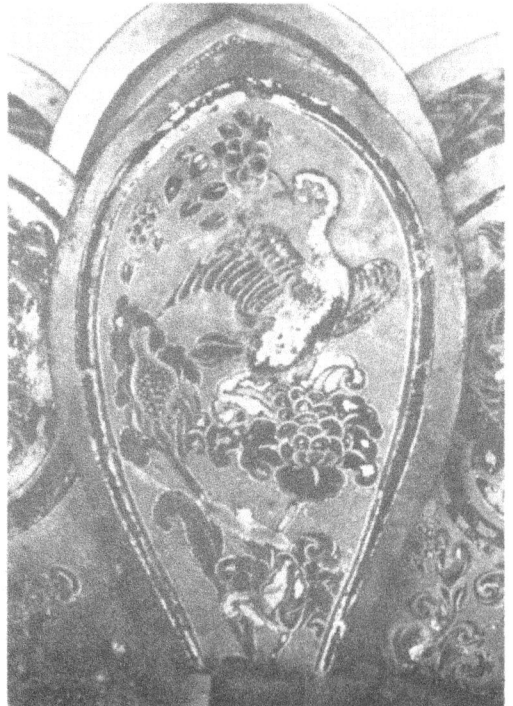

FIG. 35. Detail of painted plates of *Ganza*.

Tôdai-jin at Nara, according to the *Dai-Nippon-kobunsho gosho-shu.*[58]

Another object of related interest which is preserved in the Shōsōin, and which is illustrated in figure 37, is a *Shitsu-ban* or "wooden lacquered plate." This measures 20.5 cm. in diameter, 4.1 cm. in height and the seat itself is 19.0 cm. in diameter and 1.0 cm. high. The plate is lacquered black. The only information available from the authorities about this item is that they consider that it might have been used for "an incense clock" because the groove formed

FIG. 37.   *Shitsu-ban* in the Shōsōin, shown from above.

FIG. 36.   Detail of painted plates of *Ganza.*

by the pattern on the bottom begins and ends at the same point. It is entirely conceivable that the *Shitsu-ban* was the incense pattern for the stone seat of the *kō-in-za*, the pedestal of which is now missing.

This pattern was used in the time of Emperor Shoun (A.D. 724–748 of the Nara period), but the authorities at the Shōsōin are unable to ascertain whether this particular example formed part of a *Jikoban* or of a *Jokoban.*

Figures 37 and 38 show the *Shitsu-ban* from the top and side, and figure 34 illustrates the *kō-in-za* from

above, placed on the base alone. Figures 35 and 36 are close-up views of the painted plate-like leaves.

The *kō-in-za* and the *Shitsu-ban* from the collections of the Shōsōin are illustrated with the gracious permission of Mr. Gun'ichi Wada, chief secretary of the Imperial Treasury, who furnished the photographs.[59]

GENERAL USE OF INCENSE TIMEKEEPERS IN JAPAN

According to Professor Ruji Yamaguchi and other authorities,[60] the use of incense timekeepers in Japan is limited exclusively to Buddhist temples, and they are never to be found in the home. The one exception is the special type employed in geisha houses previously described. Professor Yamaguchi stated that the various forms of the *koban-dokei* or "incense

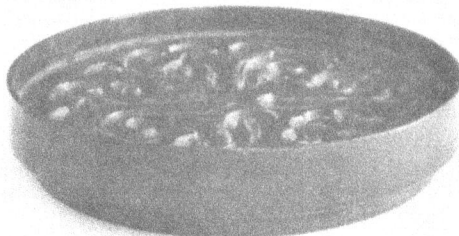

FIG. 38.   *Shitsu-ban* shown from the side.

[59] The valuable assistance of Mr. Gun'ichi Wada in this phase of the present study is gratefully acknowledged.
[60] Personal communications with Professor Yamaguchi, Mr. Howard and Mr. Blum.

[58] The kō-in-za and Shitsu-ban from the Shōsōin at Nara are also illustrated in the *Shōsōin Gyobutsu Zuroku* 11.

clocks" continued in use during the Tokugawa period (1603–1867) in the temples as timekeepers to mark the intervals at which the priests struck the great bell to call the people in the countryside to prayer. He added that although the incense timekeepers are still employed in many rural temples even now, the temple priests are not aware of their original timekeeping function and use them merely as incense burners.

The habit of prayer at certain intervals designated by the striking of the great bell of the Buddhist temples is clearly described in a passage from Golownin's work previously noted:

> The prayers are repeated three times in a day; at daybreak; two hours before noon; and before sunset: as the matin, noon and vesper masses are performed with us. The people are informed of the hours of prayer by the ringing of a bell. Their method of ringing is as follows: after the first stroke of a bell, a minute elapses; then comes the second stroke; the third succeeds rather quicker, the fourth quicker still: then come some strokes in quick succession; after a lapse of two minutes, all is repeated in the same order; in two minutes more, for the third time, and then it ends. Before the temples, there stand basins of water, made of stone or metal, in which the Japanese wash their hands before they enter. Before the images of the saints lights are kept burning, made of train oil, and the bituminous juice of a tree which grows in the southern and middle parts of Niphon (sic).

Interestingly enough, no descriptions of incense timekeepers appear in the letters and papers of the great Jesuit missionary, Father Francis Xavier, who introduced the first mechanical clocks into Japan, although he noted the use of incense itself.

FATHER TÇUZU'S ACCOUNT OF INCENSE TIMEKEEPERS

An examination of the unpublished "Historia Da Igreja Do Japão" [61] of João Rodrigues Tçuzu, S.J., reveals a passage of considerable significance. Father Rodrigues went to Japan from his native Portugal in 1576 at the age of sixteen. He became so proficient in the Japanese language that he earned the title "Tçuzu" meaning "interpreter" to distinguish him from another missionary named Joao Rodrigues of a later period. His skill as a linguist brought Tçuzu to the attentions of the military leaders, Hideyoshi and Ieyasu. Tçuzu remained in Japan for forty-five years, until his death in Macao in 1633.

In his manuscript *Historia*, Father Tçuzu wrote:

> In Japan there are no ordinary clocks to keep time, but the bonzes have certain clocks of fire in their temples to keep them informed as to the hours of prayer and to strike the hours. These clocks, which are very ingenious devices, keep time on both long and short days by means of measures which are determined according to the length of the day, which is always divided into six hours, whether the day is short or long. The clock is made as follows: it has a square wooden case that is filled with a certain kind of very finely sifted ash, which is also very dry, and

the surface of which is very flat. Using a certain measure that they have, they make in the case some furrows of a certain length, breadth and depth in the form of a square. These furrows are filled with a certain powder or flour made from the bark of a particular tree, which is very dry and aromatic. At the beginning of one of the furrows a fire is lighted, which burns very slowly so that it takes an hour to reach the end of the furrow, and the hour is measured very well, for they go by their previous experience in letting the fire burn, and it always burns in the same way and at the same rate, but the furrow is longer or shorter according to the length of the day or the night. The Chinese have sundials. They also have water clocks with which they tell time. . . .

This account of Japanese incense timekeepers written more than three hundred years ago is a succinct and accurate description, and summarizes almost everything about these implements that is known. Tçuzu pointed out, for instance, that the timekeepers were used specifically and exclusively by the bonzes in the Buddhist temples, and not by the general public. It is apparent that the incense timepieces did not undergo any changes in construction over a considerable period of time.

FRAISSINET'S ACCOUNT

The incense-seal timekeepers of Japan were noted by Western travelers as late as the nineteenth century. Edouard Fraissinet [62] in his work on contemporary Japan included a chapter on Japanese timepieces which described incense clocks as follows:

> Formerly, that is to say in very ancient times, use was made in this country (Japan) of ignescent clocks (*horloges ignées*). These were maintained by means of a fulminant powder extracted from the bark of *Illicium religiosum*. The guardians of the clock sprinkled this substance in a series of furrows laid out on a bed of ashes; and it was the progressive burning of the ignited powder that indicated to the guardians the passage of the hours which they were required to announce by the ringing of the bell. In order to regulate the movement as much as possible, the entire device was enclosed within a box ventilated only by a limited number of holes. But, towards the middle of the seventh century, this ingenious procedure was replaced by clepsydrae (water clocks) which were recognized to be more exact.

It has been noted previously in this study that the *Illicium religiosum*, known to the Japanese as the "shikimi-tree" had special significance in Buddhist religious rites, and even to the present time tree branches from this tree are placed in bamboo cups on one of the altar shelves as a necessary part of the offerings at the household *butsudan* or altar.

The statement made by Fraissinet that the incense clock was displaced by the introduction of clepsydrae in the seventh century is of particular significance. Inasmuch as the incense timekeeper is still to be found in temples in remote areas to the present time, and since it was very definitely in use in Fraissinet's time,

[61] P. João Rodrigues Tçuzu, S.J., "*Historia da Igreja do Japão*" *Colecção Noticias de Macau* (Macao 1956) 2, c. 15, pp. 129–130.

[62] Edouard Fraissinet, *Le Japon Contemporain* (Paris, 1857), pp. 170–171.

the implication is that it had a much greater distribution before the seventh century. It is interesting to note that Buddhism, which was introduced into Japan in the fifth century, received Imperial recognition and became the State religion of Japan in the reign of Empress Suiko (A.D. 593–628), in the seventh century.

INCENSE TIMEKEEPERS IN PRIVATE COLLECTIONS

Very few examples of incense clocks or *koban-dokei* have found their way into collections, public or private. The main reason may be that they are relatively unknown as timepieces to the collectors,

FIG. 41. *Jokoban* from the collection of Mr. Ed S. Jones.

but it is also true that incense clocks in any form are rarities.

Figures 39 and 40 illustrate two *koban-dokei* from the famous collection of the late Mr. N. H. N. Mody[63] which was destroyed during World War II. The first example is described in the Mody Catalogue as being "*Red Lacquer Koban-Dokei*, (Incense Clock) with lattice-work cover, and two drawers containing metal utensils for the 'clock'." The second timepiece, shown in figure 40, is a

Black Lacquer Koban-Dokei (Incense Clock) with drawer containing utensils for the clock. The incense box was half covered with ashes and then divided into compartments with incense-tablets marked with the Chinese Zodiacal Signs. The tablets were lighted at one end, and, as they slowly burned, they told the time. This method of time reading is still practiced in old temples, particularly in connection with the reading of *Sutras*.

[63] N. H. N. Mody, *A Collection of Japanese Clocks* (Tokyo, 1932), plate 114 and facing page.

FIGS. 39 & 40. Two *koban-dokei*, reproduced from Mody.

The two examples in the Mody collection are consistent in all details with the standard form of the incense seal in Japan, and the minimum information provided points up the fact that these timepieces are relatively little known even among the Japanese.

Figure 41 illustrates a *koban-dokei* with an intricately carved exterior, with motifs of clouds and scrolls.[64] The corners are reinforced with small brass corner plates. The interior is fitted with a tray made entirely of wood, having a continuous channel consisting of seven parallel lines connected at the ends. Each length is presumably for the duration of one time period. The underpart of the case and the base

[64] From the collection of Mr. Edward S. Jones, Los Angeles.

are constructed from a number of separate pieces of wood each intricately carved and fitted together. The assembly is held in place by means of a metal loop which projects into the bottom of the interior of the case, and fastened with a wooden peg. When the peg is removed, the entire base and underpart of the case disassemble into numerous units.

This timepiece, which appears to be of the eighteenth century, is made of oak (*kashi*) on the outside and lined within with pine (*kiri*). The four main panels of the exterior are representative of clouds (*yun*) and waves (*po lan*) on opposite sides, and of the lotus and the artemisia on the two remaining sides. The lower panels are carved with the symbolic motifs of waves and of mountains on opposite sides, and with a diapered design of the earth or land symbol on the other. It measures $12\frac{3}{4}$ inches square by 12 inches high.

Another incense timepiece is shown in figure 42.[65] Although otherwise similar to the example in the Mody collection, it has no pedestal. It is made of the same unusual hardwood with dominant grain, which may be either cryptomeria or hinoki cypress. The exterior is finished in clear lacquer except for the edges, cover, and turntable, which are painted with black lacquer. The case is $7\frac{1}{2}$ inches square and 12 inches high. Each section is assembled with wooden pegs and the only metal used is in the pulls for the drawers. The *koban dokei* consists of six separate parts: a latticed cover, the ash receptacle with a round

[65] From the collection of the writer.

FIG. 42. *Jokoban* from writer's collection, showing parts and utensils.

hole in the center of its bottom, a turntable consisting of a platform with ornamental skirt and a dowel in the center into which the ash receptacle fits, a chest with two drawers for storage.

The bed of powdered ashes with which the receptacle is filled is about two inches in depth. The utensils are also carved of wood. The ashes are leveled by a wooden implement (a), and a rudimentary rake is used to prepare them (b). The incense pattern consists of a wooden tray (c) with a raised edge on two sides and a groove underneath. This tray is fitted into each of the four corners of the ash receptacle progressively by means of the groove, which sets into the edge of the ash receptacle. A wooden tamper (d) is inserted along the extent of the channels in the pattern to make an impression into the ashes. Powdered incense is then raked into the channels by means of the comblike implement. The pattern tray is turned clockwise on its pivot as the incense pattern is laid so that the next corner is in readiness for the same treatment. When the incense trail has been laid on all four sides, it is noted that the pattern of the completed incense trail is in the form of a suavastika with double endings. This is a Buddhist symbol denoting "the heart of Buddha." [66]

After the incense trail has been laid, little bamboo pegs inscribed with the characters for the time intervals are inserted along the trail to divide it. The twelve zodiacal signs or *Junishi* are used as horary symbols for the hours 1 through 12. The incense trail is ignited at the starting point and it burns continuously until the end is reached.

In the few examples of the *koban-dokei* which have found their way into collections, none have retained a set of the markers for the time intervals. The constant scorching to which they would be exposed undoubtedly required frequent replacement.

An unusual form of the *koban dokei* was reported by Mr. Edwin Pugsley during a visit to Japan.[67] Among the examples on exhibit in the National Science Museum, he noted a later form of incense clock with a removable cover having little chimneys along the route of the incense trail so that the progressive passage of time could be noted from the issuance of smoke from one of the chimneys.

SMELLING THE TIME

Another aspect of the incense clocks in Japan is the possible use of several incense recipes for indicating

[66] The Suavastika or Fylfot is widely used as a magical charm. It is one of the 40,000 characters of the Chinese written language and signifies *Wan* or "10,000" and is supposed to be in itself an accumulation of 10,000 felicities. *Wan* is one of the two characters of the well-known Japanese slogan *ban zai* which is a vociferous spell uttered by loyal persons wishing their ruler all earthly happiness. It is ordinarily accepted as the accumulation of lucky signs possessing 10,000 virtues, being one of the 65 mystical figures which are believed to be traceable in the famous footprints of Buddha.

[67] Personal communication with Mr. Edwin Pugsley.

the time intervals. It is entirely possible that the pegs or tablets marked with the zodiacal characters inserted along the incense trail to denote time intervals could in actuality be tablets of hard-paste incense, each made from a different recipe. When the progressive burning of the path reached one of the markers, the marker too would be consumed and the resulting variance in aroma would be detected by the priest in attendance, so that he could tell the particular hour which had elapsed from the scent.

INCENSE TIMEKEEPERS IN KOREA

Although no examples of the incense seals or more elaborate forms of incense timekeepers appear to have been used, some of the simpler types of incense clocks were known in Korea.

It is a matter of interest to note that Buddhism was introduced into Japan from Korea. Consequently, the use of incense, which invariably accompanied the spread of Buddhism, was known at an earlier period in Korea than in Japan.

Buddhism (*Pul-to*) became the state religion of Korea during the period between the fourth to the fourteenth century, and it has survived in its Chinese form to the present. It had its greatest period in the seventh and eighth centuries A.D., when many priests migrated from China of the T'ang dynasty, and many Buddhist temples were built in Korea. It is quite certain that incense was used in an early period of the Korean dynasty, from about the tenth century A.D. Surviving records indicate that a variety of types of incense were imported from China, and that numerous other recipes were developed in Korea.

Hamel and Griffis[68] reported that incense sticks or "joss perfumery" were very much in vogue and formed part of the Buddhist ritual in modern Korea. An interesting use of incense was made by the Buddhist bonzes in the ceremony of *pul-tatta*, or "receiving the fire." This is the rite undergone by young men when making the vows of Buddhist priesthood. A *moxa* or cone of burning incense was laid upon the arm of the professing young man, after the hair has been shaved off. The cone was then ignited and it burned slowly into the flesh, leaving a painful sore, the scar of which served as a mark of dedication and holiness, and record of the initiation. However, if the professing vows were later broken, the torture was repeated on each occasion, and thus ecclesiastical discipline was maintained.

Incense is used also in the Korean cult of ancestor worship, which is almost identical to that practiced by the Chinese. Every prosperous home includes a household altar with the gilt and black tablets inscribed with the names of the departed. Sacrifices

[68] William Elliot Griffis, *Corea, The Hermit Nation* (New York, 1902); and Hendrik Hamel, "An Account of the Shipwreck of a Dutch Vessel . . . ," *Transactions of the Royal Asiatic Society* 9: 1918.

FIG. 43.   Korean incense sticks in burner, Yi dynasty, sixteenth century.   (Photo courtesy Dr. Sang Woon Jeon.)

are placed before them daily, and incense burns on the altars every day.

The use of incense for timetelling in Korea appears to have been introduced in the later Silla dynasty, in approximately the seventh and eighth centuries A.D. Incense burners of this period have been recovered in the Kyung-joo area. No special forms for incense burners used for timetelling purposes in the temples appear to have been derived. The manner of their application appears to have been quite similar to that practiced in China.

The metal incense burner, which was usually made of bronze or brass, was filled to below the rim with finely powdered ash. Into this was inserted a number of incense sticks, which the Koreans called *man-soo-hyang*. As the sticks burned, it was possible to tell time from the divisions with which they were marked (fig. 43).

According to Dr. Sang Woon Jeon of Seoul, in the past incense burners were employed for timetelling purposes only in the larger Buddhist temples, such as Ro-Jeon. The water clock of the community was established and maintained in the larger temples.

Another device for timetelling, which was used in smaller Buddhist temples of Korea, was a lamp called an *Ok-dungjan*, in which a special type of vegetable oil was burned for this purpose (fig. 44).

INCENSE TIMEKEEPERS OF NORTHEAST ASIA

The use of incense for time measurement has not been noted in Manchuria, Mongolia, or Tibet, although incense for ceremonial purposes is widely employed in these countries. On this basis it may be presumed that some form of the incense seal has been used in northeast Asia.

Some basis for this presumption may be found in an unusual incense burner illustrated in figure 45.[69] This utensil is made of copper with a brass cover. It measures $5\frac{1}{2}$ inches in height, 15 inches in length, and 3 inches in breadth. Brass bosses executed in repoussé are affixed to both sides of the case by means of copper rivets. On one side two cloud dragons face each other with the pearl of wisdom between them. On the other side are the Buddhist symbols of the canopy, the twin fish, the sacred vase, the lotus, the conch shell, the sacred knot, the umbrella, and the wheel. An elaborate grotesque mask is attached to each end, and the brass cover perforated with a floral design in repoussé enables the smoke to be emitted. The cover is surmounted with the symbol of the lotiform Wheel of the Law flanked by two antelopes kneeling on lotus petals. In spite of its overabundance of ornamentation, the receptacle was carelessly constructed and is evidently of northeastern Asiatic origin of the nineteenth century. It was acquired in Outer Mongolia but may be Tibetan or Chinese handiwork.

The burner is not fitted with an incense pattern tray, nor does it have the customary utensils for the

FIG. 44.   Korean ok-dunzjan, Yi dynasty.   (Photo courtesy Dr. Sang Woon Jeon.)

[69] From the collection of the Newark Museum, illustrated with permission of the Curator of the Tibetan Collection.

FIG. 45.　Incense burner and/or timepiece from the Tibetan collection of the Newark Museum.

incense timekeepers, but it is quite likely that it may have been used for that purpose. According to the description:[70]

When used as an incense burner this long, deep traylike receptacle was filled with sifted wood ash on which were placed glowing bits of charcoal. Pellets of incense were then laid on the glowing coal. The ash served as an insulator so that the burner never became overheated.

FIRE AND INCENSE CLOCKS IN HAWAII

The use of fire and incense for time measurement is to be found occasionally in remote areas to which the Chinese and/or Japanese have migrated, such as, for instance, among the Chinese and Japanese inhabitants of the Hawaiian Islands.[71]

A number of Oriental families on the Islands are familiar with the metal incense-seal clock of China and examples are to be found occasionally even at present as part of the furniture of the household altar, where they now serve as incense burners. Most of the owners are not aware of the timetelling function of the devices, however. One *haole* (white woman) remembered that the device was used in her childhood home for the burning of a certain type of incense to keep the mosquitoes away.

Another use of fire for time measurement in Hawaii is to be noted in this connection. Although the measurement of time is, and has always been, a matter of minor concern to the native Hawaiians, they nevertheless evolved a crude form of timepiece. They strung the kernels of kuikui nuts (also called "candle nuts") on cocoanut ribs and used them as

candles. It was always the chore of the youngest child in the household to attend the device. When a nut was almost completely burned, the child was to turn the device upside down long enough for the next nut to ignite.

Inasmuch as the use of incense in Buddhist worship originated in India, it is entirely likely that the box type of incense timekeeper used in China and Japan was originally derived from a similar form which originated in India, and that it was introduced into China at the same time as the Buddhist religion.

Of all the many methods of time measurement devised by man, the employment of fire remains one of the most intriguing for it involved the employment of yet another of his senses. The evolution of time measurement involved first the sense of sight and then the sense of hearing. In the Western World, the sense of touch was utilized with watches for the blind which were produced as early as the seventeenth century. The incense seal and other forms of incense timekeepers required the use of still another sense, the sense of smell. It is typical of the high development of Oriental culture that in addition to the senses of sight and hearing and touch, it harnessed the sense of smell for the Scent of Time.

In the burning of incense we see that so long as any incense remains, so long does the burning continue, and the smoke mount skyward. Now the breath of this body of ours—this impermanent combination of Earth, Water, Air and Fire—is like the smoke. And the changing of the incense into cold ashes when the flame expires is an emblem of the changing of our bodies into ashes when our funeral pyres have burned themselves out.[72]

[70] *Catalogue of Northeastern Asiatic Art* (Toledo, 1942).
[71] Personal correspondence with Mrs. H. Ivan Rainwater.
[72] From Kujikkajo's "Ninety Articles," quoted by Myōden.

APPENDICES

A. HORARY SYSTEMS

Three basic horary systems existed in China from the most ancient times. Reckoning was based on "the natural day" which began at sunset and the division of which varied with the seasons.

One of the earliest of the systems, which appears to have been original with the Chinese, divided a complete day from midnight to midnight, into 100 *k'o* or quarters, each of which was equivalent in modern European timetelling to 14 minutes and 24 seconds. The number of divisions was originally 100, and efforts were later made to revise this total to either 96 or 120 to conform with the duodecimal and other methods of time division.[73]

A second system divided a complete day from midnight to midnight into 12 double-hours or *shih*. The *shih* were often divided once again into halves, the first of which was called the *ch'u* and the second of which would be the *ch'eng*. The first *shih* straddled midnight so that the *ch'u* was the period from 11:00 P.M. to 12:00 midnight in European reckoning, and the *ch'eng* was the period from 12:00 midnight to 1:00 A.M. The system of 12 double-hours may have been introduced into China from Babylonia, where it was known from the earliest times. The Chinese named each of the *shih* for one of the signs of the Zodiac in the following order:

11:00–1:00 A.M.	*Tzu* (rat)
1:00–3:00 A.M.	*Ch'ou* (ox)
3:00–5:00 A.M.	*Yin* (tiger)
5:00–7:00 A.M.	*Mao* (hare)
7:00–9:00 A.M.	*Ch'en* (dragon)
9:00–11:00 A.M.	*Ssu* (snake)
11:00 A.M.–1:00 P.M.	*Wu* (horse)
1:00–3:00 P.M.	*Wei* (sheep)
3:00–5:00 P.M.	*Shen* (monkey)
5:00–7:00 P.M.	*Yu* (cock)
7:00–9:00 P.M.	*Hsu* (dog)
9:00–11:00 P.M.	*Hai* (boar)

The system of 12 double-hours was well known in China during the Han period (202 B.C.–A.D. 9) but it is believed to have had an even earlier origin.

The third horary system consisted of "night watches" by means of which the night from sunset to sunrise was divided into five equal *keng* or "watches." In order of their appearance they were known as

jih ju (sunset)
hun (dusk)
ch'u keng (10 *k'o* after dusk)
tai tan (period of "waiting for dawn")
hsaio (dawn).

---

[73] Needham, Ling & Price, *op. cit.*, pp. 199 et seq.

In his travels Marco Polo[74] noted that in every Chinese city there were stone towers to which the inhabitants might remove their effects for security in case of a fire, which were apparently frequent. By imperial edict a guard of ten watchmen was stationed under cover, five on duty through the day and five during the night:

Each of the guard-rooms is provided with a sonorous wooden instrument as well as one of metal, together with a clepsydra (*horiuolo*) by means of which latter the hours are ascertained. As soon as the first hour of the night has expired, one of the watchmen gives a single stroke upon the wooden instrument, and also upon the metal gong (*bacino*) which announces to the people of the neighboring streets that it is the first hour. At the expiration of the second, two strokes are given, and so on progressively, increasing the number of strokes as the hours advance. . . .

Japanese timetelling is based on the Chinese system of "double-hours" varying with the seasons, with the exception that the natural day begins at dusk instead of at sunset. The Chinese zodiacal signs were adopted, but whereas in China the sign represented the middle point of the *shih*, the Japanese reckoned the beginning of the hour. As an example, while the "Tiger" hour was the period from 3:00 to 5:00 in the morning, in Japan it designated the period from 4:00 to 6:00.

According to a study of the calendar and time measurement published by Hirayama Kiyotsugu,[75] the reckoning of time was first developed in Japan in about A.D. 660 by Prince Tenchi, who later became emperor. The natural day was divided into 50 *koku*, in accordance with the Guchu calendar. One *koku* consisted of 6 *Bu*. The day was divided into twelve double-hours or *Shinkoku*, each of which consisted of 4 *koku* and 1 *Bu*. This division was generally used by the public but the author specified that a more accurate division of a day into 100 *koku* was used by the astronomers and calendar makers. By this method 1 *koku* consisted of 84 *Bu*, and 1 *Shinkoku* or double-hour was made up of 8 *koku* and 28 *Bu*.

Another Japanese time system was called *Rokuji*, which meant literally "6 times" (e.g., *Roku*—6 and *ji* —times). By this system each day was divided into 6 periods as follows:

Shincho (morning)
Nichu (afternoon)
Nichibotsy (evening)
Shoya (early night)
Chuya (midnight) or *Hanya* (half night)
Koya (after night).

---

[74] (Marco Polo), *Travels of Marco Polo* (London, 1925).
[75] Hirayama Kiyotsugu, *Rehiko Oyobi Jiho.*

According to Yabuuti, T. Oda stated that this time system was based on the practice of praying to Amitabha at six intervals throughout the day, a practice which was known as *Rokuji no raisan* which may be translated literally as "admire [Amitabha] six times."

B. HISTORY OF INCENSE

Inasmuch as incense plays a critical role in the present study, it is necessary to note briefly its origin and use in the countries of the Orient.

Incense is defined to be both the perfume or fumigation arising from the burning of certain resins, barks, woods, seeds, fruit, etc., as well as the material being burned itself.

Its origin is lost in the mists of antiquity, but the history of its use is laden with an infinity of tradition and detail. It is presumed that incense was first used in the countries of the Near East, and that it was next introduced into India. From India its use spread simultaneously with the religion of Buddhism into Nepal, Tibet, Ceylon, Burma, China, Korea, and Japan. In Egypt incense was used as early as Dynasty XVIII (1580–1350 B.C.), and in India it was known from the most ancient times according to references which occur in the classic *Ramayana* and *Mahabharata*.

The number of substances which have been used for incense is endless. Perhaps the best known is frankincense which is derived from the Indian tree *Boswellia thurifera*. The tree was indigenous in Arabia and Africa, from which it was brought into India. From India it was exported into the countries in which Buddhism was introduced. A common incense used in India in modern times is benzoin. Incense sticks sold throughout the country are called *ud-buti* or "benzoin lights" as well as *aggar-ki-buti* or "wood aloes lights."

Probably the most popular material for making incense in the Orient was sandalwood (*Santalum album*). This is a fragrant wood native to India, where its use has been traced to at least the fifth century B.C. Tradition relates that the first image of Buddha imported into China from northern India soon after the Christian era by the earliest Buddhist pilgrims was carved of white sandalwood. The wood is still being imported from India into China for the express purpose of carving into Buddhist images. It is also used in the form of chips, which are piled before the Buddhist altars and ignited. Sandalwood is prepared also in pulverized form for molding into perfumed incense sticks to be burned before sacred shrines. It is a wood favored for carving the rosaries worn by Buddhist monks and in general it is used wherever Buddhism exists. The archaic name for the wood in China was *chantan* and in modern Chinese it

is known as *tan-hsiang* or "sandal perfume" or "sandal incense."[76]

Special dictionaries and manuals were published in China and Japan for the classification of the varieties of incense and recipes for preparing them. Special incenses were reserved in princely families, and recipes were handed down from one generation to another. The variety of materials and mixtures was infinite. Ashikaga Yoshimasa, who wrote on the subject,[77] collected and named 130 varieties of mixed incense.

The uses of incense varied in the countries of the Orient. In China, for instance, it was employed primarily for religious purposes, and formed part of the ancestor worship, the Confucian ritual and in the ceremonies of the Buddhist religion. In Japan, however, the only religious use of incense was in connection with Buddhism.

The prevalent use of incense for religious ceremonies was noted again and again by European travelers in the Orient. One of the great travelers was Marco Polo in the thirteenth century, who furnished one of the first records of incense in the Orient.[78] In his account of a visit to the palace of the great khan at Shandu he described the position of the *Baksi*, a religious order from Tebeth and Kesmir, who frequented the palace. When the festival days of their idols drew near, they asked the great khan for appeasement of their idols in the form of

. . . a certain number of sheep, with black heads, together with so many pounds of incense and of lignum aloes, in order that we may be enabled to perform the customary rites with due solemnity.

Marco Polo went on to relate that the kingdom of Kanan or Tana

. . . produces a sort of incense, in large quantities, which is not white, but on the contrary of a dark colour. Many ships frequent the place in order to load this drug. . . .

The "drug" was probably benzoin, which was not produced in India, but was imported from Sumatra to supply the markets of Arabia, Persia, Syria, and Asia Minor. In his account Marco Polo noted that the Tartars were idolators, and that they burned incense before certain tablets inscribed with names, and before an image of Natigai.

Another famous traveler, Friar Odoricus, related in 1330–1331 that when the Great Khan traveled through any country, the subjects kindled fires before their doors, casting spices thereinto to make a perfume.[79]

Friar Manrique, the Augustinian missionary who voyaged into Portugese Asia in the seventeenth century, described a festival at a temple on the island

[76] Kiyoshi Yabuuti, "Chugoku No Tokei," *Japanese Journal of the History of Science* 19 (1951).
[77] Stephen W. Bushell, *Chinese Art* (London, 1906).
[78] Lafcadio Hearn, *In Ghostly Japan* (Boston, 1890).
[79] (Friar Odoricus), *Journal of Friar Odoricus* (London, 1947).

of Saugar.[80] Inside the temple the music and "hot wafts of scent from flowers or incense worked on minds open to bewitchment."

Matteo Ricci, the founder of the Jesuit mission in China in the seventeenth century, described the memorial shrines built to honor magistrates when they left the cities over which they had presided, and he noted that

A yearly allowance is granted for these temples for incense and to pay servants for attending to perpetual lights. The large incense bowl, made of bronze, used in these shrines is similar to that employed for the same rite in the veneration of the idols. . . .

Ricci also described the use of incense offerings in the seasonal worship at the Temple of Confucius, as well as in the Confucian temples honoring titular spirits of the cities.

In Japan, Saint Francis Xavier made little note of the use of incense but he encountered it on his voyage aboard a Chinese ship bound from China to Japan over the South China Sea.[81] In a letter to the Jesuits at Goa he described a Chinese sea-god's shrine aboard the junk and wrote:

. . . the heathen confiding in that idol of theirs on the poop, which they worshipped with lighted candles and incensed by burning before it sticks of sweet-smelling wood, while we put our trust in God the Creator of Heaven and Earth. . . .

The employment of incense was mentioned by later travelers and some of these accounts provided additional details. Stoddard[82] stated that in the Temple of Five Hundred Gods at Canton the worshipers burned a stick of incense in a small jar of ashes before each of the idols. Sven Hedin[83] noted that in the Temple of Ta-fo-szu at Jehol "a lama stepped forward, bowed and lighted sticks of incense in a bronze bowl half full of ashes." The same use of a bowl of ashes was mentioned in Japan by Lafcadio Hearn in a story in which he described a shrine adjacent to a Buddhist temple which he saw "through the blue smoke that curls up from half a dozen tiny rods planted in a small brazier full of ashes." [84]

The Marquess Curzon of Kedleston observed, when he visited the Buddhist monastery of Ku-shan on Drum Mountain, that below the images of the three Buddhas there arose from the altars ". . . a thin smoke curling upwards from the slow combustion of blocks of sandal-wood, or from sheaves of smouldering joss-sticks standing in a vase." [85]

The relation between incense and Buddhism has been constantly noted. Hearn described the statues of the Roku-Jizo in a Buddhist cemetery. Each held a symbolic object, such as a lotus, a pilgrim's staff, and a Buddhist incense box among other items. In his account of a pilgrimage to Enoshima, Hearn reported that "at the right hand of Shaka enthroned on a lotus some white mysterious figure stands, holding an incense-box. . . ."

According to such writers as Hearn,[86] Clive Holland,[87] and Okakura-Yoshisaburo,[88] every Japanese household of the better or the so-called old-fashioned class included an altar which formed the center of the family's religious life. Since Shinto was the national religion in Japan, every home had a Shinto altar. If the family was Buddhist, every household had a second, Buddhist, altar as well.

The Shinto altar was in the form of a plain, wooden shelf called a "Kami-dana" or "God's shelf" made somewhat in the form of a Shinto temple or "Mitayama." This was placed over a doorway. However, incense played no part in the Shinto ritual and was not included among the offerings.

The Buddhist altar or "butsudan" in the home was invariably set against the outside wall of the inner back room. It was constructed in the form of a miniature Buddhist shrine in which an image of Buddha was housed, as well as the "ihai" or "soul commemoratives." These were mortuary tablets dedicated to the memory of deceased members of the family. A lamplet, an incense cup, and a water vessel were part of the furniture placed before the shrine, as well as candles, bundles of incense sticks, and perhaps a pair of bamboo cups containing sprays of the sacred plant Shikimi (*Illicium religiosum*). The incense was burned before the altar in the daytime, while at night a little floating wick lamp was kept alight. Taylor[89] noted that before the image of Buddha on the butsudan, ". . . in the porcelain bowl of ashes, stand glowing bundles of fragrant wood, irreverently styled 'joss-sticks' by the infidel, which waft little clouds of incense before great Buddha."

Three classes of incense were used in the Buddhist rites. First of all was the *kō*, which is the Japanese word to denote hard incense or incense pastilles, literally "a fragrant substance." This type of incense, which is produced in great variety, is burned. The second form is the *dzukō* or an odoriferous ointment which is rubbed on the hands of the priest as an ointment of purification. Finally there was the *makkō* or "fragrant powder" which was sprinkled about the sanctuary. This powder is stated to be identical with powdered sandalwood which is very frequently

[80] Maurice Collis, *The Land of the Great Image* (New York, 1958).
[81] James Broderick, S.J., *Saint Francis Xavier 1506–1552* (London, 1958).
[82] John L. Stoddard, *Stoddard's Lectures* (Chicago, 1925) 3.
[83] Sven Hedin, *Jehol, City of Emperors* (London, 1932).
[84] Lafcadio Hearn, "Jizo," *A Japanese Miscellany* (Boston, 1908).
[85] Marquess Curzon of Kedleston, *Leaves from a Viceroy's Notebook* (London, 1927).

[86] Lafcadio Hearn, *Glimpses of Unfamiliar Japan* (Boston, 1895).
[87] Clive Holland, *Old and New Japan* (London, 1907).
[88] Okakura-Yoshisaburo, *The Life and Thought of Japan* (London, 1913).
[89] Bayard Taylor, *Japan In Our Day* (New York, 1893).

mentioned in Buddhist texts. To the Buddhists, incense is "the Messenger of Earnest Desire" and symbolizes the Pious aspirations of the faithful.

The use of incense was influenced by other religious beliefs which originated prior to the establishment of a formal religion. Incense is still burned, for instance, in the presence of a corpse so that the fragrance will shield the body and the liberated soil from evil demons. It is also used to summon spirits or the souls of the absent and for this purpose the incense is called *Kwan-hwan-hiang* or *Hangon-kō*. According to Japanese legend, any form of burning incense will summon *Jiki-kō-ki* or "incense-eating goblins," a class of *pretas* recognized in Buddhism. These are the souls of men anciently guilty of selling bad incense for gain and that are now compelled to seek their only food in incense smoke.

In the work by Beardsley-Hall-Ward[90] the use of incense is noted in connection with memorial services for Buddhist dead which occur four times a year. The ancestral tablets are exposed with an incense bowl before them, a candle and a jar of shikimi leaves flanking food offerings. The Shikimi trees are considered to be sacred plants among the Buddhists, and not only are branches of the tree displayed on the butsudan, but the bark of the tree is used as an ingredient for making incense.

Incense forms part of the offering made on the occasion of the funeral of a deceased Buddhist in Japan. Before a temporary tablet on which the name of the deceased is inscribed, a Buddhist priest delivers this customary incantation:

In my heart's core I respectfully request that the scent of this stick of incense offering from the heart may pervade the regions where the Law prevails and that the Messenger of Hades may conduct the soul thither.

A fitting commentary on the modern religious uses of incense in the Orient was expressed by Lew Ayres:[91]

Cautiously used, incense can contribute greatly to a highly satisfying liturgical experience. Excessively used, it simply becomes a revolting, irritating stench. Throughout the Orient incense is used with utter abandon. One's last few coins are spent for it. In many places it has become the custom to use not fewer than three sticks at a time. It is burned profusely day and night. The temples reek of it. One wonders if there is any significance in the fact that since ages past the temple owners have been the leading incense concessionaires.

Buddhism reached Tibet in the seventh century A.D. and the religion was spread by priests from India, and developed into a form that was peculiarly Tibetan. Its priests or "lamas" multiplied rapidly and overtook the authority of the country in matters of state until the kingship of Tibet was assumed by the priesthood in the form of the Dalai Lama.

The role which incense played in Tibet was comparable to its role in China and Japan. It is offered daily in temples and in the home. An incense burner is part of the altar furniture and it is placed on a lower shelf or in front of it. When incense is not readily available or too expensive, juniper twigs serve as a substitute. An incense kiln is to be found beside the door of almost every home, where the juniper twigs are burned every morning and every evening. Tibetan travelers place lighted incense sticks in rocky clefts where evil spirits are believed to lurk.

In the Tibetan temples the lama offers incense at dawn, in elaborate censers of gold and silver, to the several classes of divinities. Incense is among the Eight Essential Offerings which form part of every rite. On the anniversary of Buddha's death, incense is burned on every hilltop, in every temple, home shrine, and lamasery.

The same general use of incense is found again in Mongolia and in Manchuria. Roerich[92] noted that at sunset in Ladak smoke arose slowly above every house on the plain as incense was ignited during the hour of prayer. He also reported that the shops of Urga sold large stocks of objects used in temple ceremonies and that incense sticks of Tibetan and Chinese manufacture were popular items. The incense sticks made at the Sera Monastery at Lhasa were particularly prized in Mongolia.

In addition to its role in religious worship, incense plays a part in social entertainment as well. In Japan *kiki-kō* or "incense sniffing" has been an aristocratic amusement since the thirteenth century. Hearn recorded that incense parties were invented before the time of the Ashikaga shoguns and were especially popular during the Tokugawa rule. Several forms of *kō-kwai* or "incense games" existed, each with its own very rigid etiquette. The host produced three kinds of incense, which he selected from a score of varieties and the guests supplied another. These were divided into four packages of each. The guest's incense was not divided, however. The host burned a package of each incense in a burner, announced the incense by a number only, and it was passed from guest to guest who sniffed the smoke. After each package had been burned, the guests were given numbered tallies. The host again prepared incense in the burner, burning one variety at a time, without announcement. Now each time the guests, after having sniffed the incense, dropped a tally with the number he supposed to be the mark indicating the variety, into a box. After the varieties had been all sniffed and noted, the results were examined and a victor was announced. The utensils required for the game were extremely expensive and were wrought of the finest gold lacquer or precious metal, and many were rare works of art. The incense parties ranked

[90] Richard K. Beardsley, John W. Hall and Robert E. Ward, *Village Japan* (Chicago, 1959) 2.

[91] Lew Ayres, *Altars of the East* (New York, 1956).

[92] Georg N. Roerich, *Trails to Inmost Asia* (New Haven, 1931).

second only to the tea ceremony among persons of prestige and taste.[93]

The manner in which incense sticks were made in China in modern times was described by Gontran de Poncins.[94] Three-quarters of the length of the stick consisted of wood dust derived from grinding wood in stone mortars with primitive pestles. The other quarter is compressed from a combination of the bark of elm root powdered and thinned with water, to which were added scented powders of incense, clove, camphor, and odoriferous woods such as cyprus. The scented powders were thinned with Chinese wine. The elm root bark and the powders were all ground together and made into a binding paste placed in a type of pump which crushed the mass and from which it emerged by means of round holes at the end of the pump in the form of sections molded like thin wire. These were dried and then cut up into ritual lengths, usually from 6 to 7 inches. The dried sticks were then sold in batches of 19, 37, 61, or 91.

There is a very simple explanation for these apparently haphazard and mysterious quantities of incense sticks in batches of various sizes. They are merely the quantities that result from tying the sticks into cylindrical bundles. The Chinese started their bundles with a single stick in the center, six sticks ringed around it, twelve more around this cylindrical bundle, another twelve sticks in the next, and so on, as follows:

$$1 + 6 + 12 = 19$$
$$1 + 6 + 12 + 18 = 37$$
$$1 + 6 + 12 + 18 + 24 = 61$$
$$1 + 6 + 12 + 18 + 24 + 30 = 91.$$

C. THE MIZU-TORI OR SPRING FESTIVAL

Participants in the water-drawing ceremonies are Buddhist priests, civil officials, and some of the pilgrims. Eleven bonzes or priests known as the *Rengyoshu* take the individual positions of *Wajyo, Daidoshi, Shushi, Dotsukasa, Kitazashu #1, Nanzashu #1, Kitazashu #2, Nanzashu #2, Chuto #1, Gonshosekai,* and *Shosekai.* Participating civil officials have hereditary offices entitled *Shoko, Dodoshi, Kishi, Komori, Ooi,* and *Inshi.* The titles given to pilgrims participating are *Kakubuyo, Tubonebugyo,* and *Doshi.*

Dr. Asahina's account of the ceremony is invaluable for an understanding of the function of incense in the ceremony.[95] After each evening ceremony, the inner room of the Nigatsudô is swept clean. The incense, which has been burning in the Jikoban since the previous evening, is cleared away and replaced with new incense.

At 6:00 P.M. the *Daishoya* announces the time to the *Shoko* by striking the temple's great bell. *Shoko* reports the time to the *Dotsukasa* in the Sanro Shukusha and then proceeds to the Nigatsudo, where he obtains the kindling coals from the *Shosekai.* He makes a ceremony of lighting the sacred lamps in the February Hall. The fire which is burning in the sacred lamp of the *Naijin* or inner room is transferred to the *Matsu No Jin,* or slender pieces of the oil-rich part of the pine roof. The same fire is used to ignite the sacred lights in the hall. As part of the same rite, the incense trail of the Jikoban, which has been remade, is lighted from the ignited *Matsu No Jin* by the *Shosekai.* The Shoko next proceeds from the temple down to the room of the Dotsukasa, or the *Dotsukasa Shukusho,* to inform the Dotsukasa that the temple has been lighted. From there the Shoko returns to his house. The Dotsukasa sends Kakubugyo to the Nigatsudô and the latter runs up the north steps at the temple entrance with a flaming pine torch in his hand. He places the torch at the temple entrance with the words "*Jiko Shoya Made × Sun × Bu*" (*Jiko* is × *Sun* × *Bu* in length before *Shoya.* 1 *Sun* is 3.030 cm., 1 *bu* is $\frac{1}{10}$ of a *Sun*).

The Kakubugyo then returns to report the time to Dotsukasa and upon the latter's order, goes on to give the time to Shoko, who informs Daidhoya with the words as before "*Jiko Shoya Made × Sun × Bu*" and thereupon the Daishoya strikes the temple bell. The time is then about 7:00 P.M.

Upon hearing the bell, Dotsukasa sends Kakubugyo to the temple hall to call out "*Yoji No Annai!*" to Shosekai. Following orders, Kakubugyo finishes the preparation of the Naijin or inner room for the impending arrival of the Rengyoshu to the temple.

Dotsukasa again commands Kakubugyo to announce to Shosekai that "*Shussi No Annai.*" When the bell rings again, the members of the Rengyoshu enter the temple in order, with Doshi carrying a torch of green bamboo which is called a *Kaku Taimatsu.* The torch is born aloft by Doshi to the front floor of the Nigatsudô and is brandished as traditionally prescribed in the manner of a rolling wheel. The torch is formed like a basket and is somewhat more than 4 *Ken* in length (1 *Ken* is equivalent to 1.81818 m., and it weighs 20 *Kan* (1 *Kan* equals 3.75 kilograms). This type of torch is used particularly during three days of the ceremony, namely, the twelfth, thirteenth, and fourteenth of March. The handling of the torch during the ceremony requires special skill. The spectacle that is created by the sparks that fall as the torch is kept in motion resemble the petals of cherry blossoms, according to the Japanese, and they find it very exciting. Soon after all the members of the Rengyoshu have entered the temple, *Shoya No Gyoho* begins.

The *Omizutori No Gyoji* or ceremony of taking the water from the sacred well begins in the middle of

---

[93] Basil Hall Chamberlain, *Things Japanese* (London, 1902).

[94] Gontran de Poncins, *From a Chinese City* (New York, 1957).

[95] Teiichi Asahina, "On the Time Measuring Incense Board," in *Yamato Bunka Kenkyu* (Research on the Culture of Yamato) 2 (1954): 3.

Koya on the twelfth day of March, or more exactly, at about 2:00 A.M. on March 13. Five of the dignitaries participate. *Kitazashu No. 1, Chyuto No. 1, Gonshosekai,* and *Shosekai* follow *Shushi* down the stone steps to the southern part of the temple hall to scoop water from the well. The well is given the name *Wakasai*. Thereupon Doshi steps forward with a *Hasu*, a large torch about 1 *Jo* in length, in his arms. Scooping the water from the sacred well is the hereditary function of the Doshi families, and they have kept the details of this well a closely guarded secret.

Accordingly, during the three days, March 12, 13, and 14, the special event of the *Dattan No Myoho* follows the event of the after-night previously described. A large torch of pine weighing about 20 *Ken* and measuring 1 *Jo* in length is revolved, showering sparks, while at the same time the Doshi take part together with the Reisuikan and Sanzyo in scattering water which is scooped from the well.

During the ceremony of igniting the Jikoban, a *Tsuketake* or wooden spill is placed at the left side of the incense trail. The flame which ignites the incense timepiece is transferred to it by means of the spill from the sacred fire which is kept constantly burning in the *Naijin,* or inner room of the Nigatsudo. The name of the spill, *Tsuketake,* is probably a corruption of the word *Tsuketate* which literally means "spill." It is made from the wood of a tree called *Shidebo,* which grows in the region or precinct of *Todai Ji.* The *Shidebo* tree, which is so called by the personages taking part in the *Mizu Tori* ceremonies, was probably originally called the *Hideboku* for it appears as such in the archaic records of the Doshi families. The name is derived as follows: *Hi* meaning "fire," *De* meaning "present" or "appear," and *boku* meaning "wood" or "tree." Dr. Asahina reported that upon examination of a twig he found it to be *Inu Shide* and consequently the tree or wood (*boku*) of the *Shide* may also have been corrupted to the name *Shidebo.* He found two varieties of the tree at Nara, the *Inu Shide* and the *Aoshide.*

Among the groves of the *Shide* trees at Nara there were a number of large, withered trees, which were fragile and porous. Dr. Asahina assumed that the spills for the ceremonies in the Nigatsudo were cut from these trunks and dried. The Doshi produce spills from these trees by cutting portions of the trunk and dehydrating them in an earthen pan.

The *Tsuketake* or spills are employed at the beginning of the *Mizu tori*, during the *Ittokuka* which is held during the early morning of March 1. During this ceremony, *Ittoku,* who is the heir of the Inagaki family which has ministered to the *Sho Kan Non* of the Nigatsudo since the Tempyo era, ignites a spill by means of a flint. The fire from the spill is transferred to the *Jotomyo* or perpetual sacred light in the inner room of the Nigatsudô.

Although the schedule of the *Shujikai* varies somewhat from day to day, it is always arranged and timed by means of the incense timepiece, and therefore the Jikoban becomes an important factor in the water drawing ceremony. In ancient times it undoubtedly played an even more important role.

BIBLIOGRAPHY

ALEXANDER, W., and A. STREET. 1956. *Metals in the Service of Man* (Baltimore).

ASAHINA, TEIICHI. 1954. "On the Time Measuring Incense Board" in *Yamato Bunka Kenkyu* (Research on the Culture of Yamato) 2: 3.

AYRES, LEW. 1956. *Altars of the East* (New York).

BEARDSLEY, RICHARD K., JOHN W. HALL and ROBERT E. WARD. 1959. *Village Japan* (Chicago).

BRINCKLEY, CAPTAIN F., ed. 1940. *Japan Described and Illustrated by the Japanese* (Boston).

BRODERICK, JAMES, S. J. 1958. *Saint Francis Xavier 1506-1552* (London).

BUSHELL, STEPHEN W. 1906. *Chinese Art* (London).

CHAMBERLAIN, BASIL HALL. 1902. *Things Japanese* (London).

CHAMBERS, SIR WILLIAM. 1757. *Traité des Edifices, Meubles et Habits de la Chine* (Paris).

CHAPUIS, ALFRED. 1919. *La Montre Chinois* (Neuchâtel).

Chung-Shan Ta-Tzu-Tien I-Tzu Ch'ang-Pien.

COLLIS, MAURICE. 1958. *The Land of the Great Image* (New York).

Dai Kanwa Jiten.

D'ELIA, P. PASQUALE, ed. 1942-1949. *Fonte Ricciane* (3 v., Rome).

DE PONCINS, GONTRAN. 1957. *From a Chinese City* (New York).

EARLE, ALICE MORSE. 1902. *Sundials and Roses of Yesterday* (New York).

ERSKINE, WILLIAM HUGH. 1933. *Japanese Festivals and Calendar Lore* (Tokyo).

FAVRIER, A. 1900. *Pékin, Histoire et Description* (Lille).

FRAISSINET, EDOUARD. 1857. *Le Japon Contemporain* (Paris).

GALLAGHER, LOUIS J., S.J. 1953. *China in the Sixteenth Century, The Journals of Matteo Ricci 1583-1610* (New York).

GOLOWNIN, CAPT. 1824. *Memoirs of a Captivity in Japan During the Years 1811, 1812 and 1813* (London).

GRIFFIS, WILLIAM ELLIOT. 1902. *Corea, the Hermit Nation* (New York).

GUITTON, ROBERT. 1958. *Quand Sonne l'Heure* (Brive).

HAMEL, HENDRIK. 1918. "An Account of the Shipwreck of a Dutch Vessel on the Coast of the Isle of Quelpaert, Together with a Description of the Kingdom of Korea," *Transactions of the Royal Asiatic Society* 9.

HARVEY, EDWIN D. 1933. *The Mind of China* (New Haven).

HEDIN, SVEN. 1932. *Jehol, the City of Emperors* (London).

HEARN, LAFCADIO. 1890. *In Ghostly Japan* (Boston).

——. 1895. *Glimpses of Unfamiliar Japan* (Boston).

——. 1904. *Japan, An Interpretation* (New York).

——. 1908. *A Japanese Miscellany* (Boston).

HOLLAND, CLIVE. 1907. *Old and New Japan* (London).

HOMMEL, RUDOLPH P. 1937. *China at Work* (New York).

Ho-nan Ch'en Shih Hsiang-P'u (Handbook of Incense by the Ch'en Family in Honan [early fourteenth century]).

Hsiang P'u (Catalogue of Aromatics by Hung Ch'u [end of eleventh century or early twelfth century]).

Hsiang Sh'eng (Comprehensive Account of the Incense by Chou Chia-Chou [late sixteenth century]).

Hsiao Hsüeh Kan Chu. 1299.

Hsin-Tsuan Hsiang-P'u (Newly Compiled Book of Incense [also known as *Ho-nan Ch'en Shih Hsiang-P'u*]).

JÜNGER, ERNST. 1957. *Das Sanduhr Buch* (Frankfort-am-Main).

KARLGREN, B. 1950. "The Book of Documents" (Shu Ching), *Bulletin of the Museum of Far Eastern Antiquities* (Stockholm) 22: 1.

KEDLESTON, MARQUESS CURZON OF. 1927. *Leaves from a Viceroy's Note-Book* (London).

KENTARŌ, PROFESSOR YAMADA. 1957. *Tōzai Kōyaku Shi* (Tokyo).

KIYOTSUGU, HIRAYAMA. N. D. *Rekiho Oyobi Jiho* (Tokyo).

KNOX, GEORGE WILLIAM. 1905. *Japanese Life in Town and Country* (New York).

KUNITOMO, HIDEO. 1948. *Tokei No Hanashi* (Tokyo).

LEWIS, NORMAN. 1951. *A Dragon Apparent: Travels in Indo-China* (London).

Lou K'e Ching.

MAGALHAENS, P. GABRIEL de. 1668. *Nouvelle Relation de la Chine* (London).

MAGOWAN, DR. D. J. 1852. "Chinese Horology." Part IX: Communications, *Annual Report of the Commissioner of Patents for 1851* (Washington).

MODY, N. H. N. 1932. *A Collection of Japanese Clocks* (Tokyo & Kobe).

NAVARETTE, DOMINGO FERNANDEZ, O. P. 1748. "Description de la Chine." L. II. *Histoire General des Voyages* (Paris).

NEEDHAM, JOSEPH. 1959. *Science and Civilization in China* (Cambridge) 3.

NEEDHAM, JOSEPH, WANG LING and DEREK J. PRICE. 1959. *Heavenly Clockwork* (Cambridge).

(Newark Museum). 1950. *Catalogue of the Tibetan Collection, Part II* (Newark).

NIEUHOFF, JOHN. 1673. *An Embassy from the East India Company of the United Provinces to the Grand Tartar Cham, Emperor of China, Delivered by Their Excellencies Peter de Goyer & Jacob de Keyser* (London).

(ODORICUS). 1947. *The Journal of Friar Odoricus* (London).

OGAWA, TOMOZO. 1838. *Teikoku Kassoku* (Practical Method for Measurement of the Fixed Time by the European Horloge (Kobe).

OKAKURA-YOSHISABURO. 1913. *The Life and Thought of Japan* (London). *P'ei-Wen Yün-Fu.*

PLANCHON, MATTHIEU. 1899. *L"Horloge: Son Histoire Retrospective, Pittoresque et Artistique* (Paris).

(POLO, MARCO). 1925. *The Travels of Marco Polo* (London).

POUND, EZRA. 1954. *The Classic Anthology Defined by Confucius* (Cambridge).

PRÉVOST, ABBÉ, trans. N.D. *Churchill's Collection of Voyages and Travels* (London).

READ, BERNARD. 1936. *Chinese Medicinal Plants* (Peking).

RODRIGUES TÇUZU, JOĀO. 1956. "Historia da Igreja do Japāo." (Ed. por João Amaral Abranches Pinto.) *Colecção Noticias de Macau* (Macao) 2: 15.

ROERICH, GEORG N. 1931. *Trails to Inmost Asia* (New Haven).

Shih-Yüan Ts'ung-Shu.

SHŌSŌIN GYOBUTSU ZUROKU. 1928. (Tokyo).

STODDARD, JOHN L. 1925. *Stoddard's Lectures* (Chicago) 3.

TAKABAYASHI, HYŌE. 1924. *Tokei Hattatsu Shi* (Tokyo).

TAYLOR, BAYARD. 1893. *Japan in Our Day* (New York).

TIFFANY, OSMOND, JR. 1844. *The Canton Chinese, or, The Americans' Sojourn in the Celestial Empire* (N.P.).

(Toledo Museum of Art). 1942. *Northeastern Asiatic Art* (Toledo).

TRIGAULT, P. NOCOLAS. 1615. *De Christiane Expeditione Apud Sinas* (Vienna & Augsburg).

WADDELL, L. AUSTINE. 1905. *Lhasa and Its Mysteries* (New York).

WALEY, ARTHUR, ed. 1954. *The Book of Songs* (London).

WILSON, ERNEST HENRY. 1913. *A Naturalist in the Far East* (New York).

YABUUTI, KIYOSHI. 1951. "Chūgoku No Tokei" in *Kagakushi kenkyu* (*Japanese Journal of the History of Science*) 19.

YING-LIN, WANG. N.D. *Hsiao Hsüeh Kan Chu* (This work by Hsüeh Chih-Hsüan was published by Wang Ying-lin in 1299).

Zenrin Zōkisen.

GLOSSARY

	Camphor (Tree), (see Chang)	洪芻	Hung Ch'u
樟	Chang	宣州	Hsüan-chou
辰	ch'en	有熊氏	Yu-hsiung Clansman
陳敬	Ch'en Ching	杜牧	Tu Mu
沈香	Ch'en-hsiang	潤州	Jun-chou
芰香	Chien-hsiang	燕肅	Yen Su
製甲香	Chih-chia-hsiang	梓州	Tzǔ-chou
周嘉冑	Chou Chia-chou	吳僧瑞	Wu Seng-jui
降真香	Chiang-chen hsiang	杭州	Hang-chou
鄒象濬	Ch'ou Hsiang-t'an	湖州	Hu-chou
篆	chuan	婺州	Wu-chou
中齋居士	Chung-chai chü-shih	睦州	Mu-chou
	Five Night Aromatic Notch, (see Wu yeh hsiang k'o)	吳正仲	Wu Cheng-chung
	Five Night Seal-character Aromatic, (see Wu yeh chuan hsiang)	香譜	Hsiang p'u
	Five Watch Seal Notch, (see Wu keng yin k'o)	熙寧	Hsi-ning
	Greatly Expanded Seal-character Aromatic, (see Ta yen chuan hsiang)	新篆香譜	Hsin tsuan hsiang p'u
		宣州	Hsüan-chou
香柬	Hsiang ch'eng		Hundred Notch Aromatic Seal, (see Po k'o hsiang yin)
香	hsiang		Hundred Notch Seal Aromatic, (see Po k'o yin hsiang)
藿香	Huo-hsiang		Hundred Notch Seal-character Aromatic, (see Po k'o chuan hsiang)
小學紺珠	Hsiao hsueh kan chu	荏	Jen (Tree)
新篆香譜	Hsin-tsuan hsiang-p'u	刻	k'o
繼隆	Hsi'lung	廣德	Kuang-tê

47

	Mountain Pear, (see Shan-li)			Solar Periods, (see the Chart for the Five Night Seal-character Aromatic)
刻	k'o		天歷　天曆	T'ien-li　or
甘松	Kan-sung			Ting-chou Public Storehouse Seal Aromatic, (see next)
廣德	Kuang-tê		定州公庫印香	Ting-chou king k'u yin hsiang
零陵香	Ling-ling-hsiang		寸	ts'un
漏刻經	Lou k'ê ching			Twelve Branches, (see Shih erh chih)
龍腦	Lung-nao		慈利	Tzǔ-li
茅香	Mao-hsiang		吳正仲	Wu Cheng-chung
茅香	Mao-hsiang		婺州	Wu-chou
盆硝	P'en-hsiao		五更印刻	Wu keng yin k'o
丁香	Ting-hsiang		五夜篆香	Wu yeh chuan hsiang
時	shih		黃熟香	Huang-shu-hsiang
紹南	Shao-nan		零陵香	Ling-ling-hsiang
睦州	Mu-chou		藿香	Huo-hsiang
楠木	Nanmu (Tree)		土草香	T'u-ts'ao-hsiang
	notch, see k'o		大黃	Ta-huang
百刻篆香	Po k'o chuan hsiang		檀香	T'an-hsiang
百刻香印	Po k'o hsiang yin		定州	Ting-chou
百刻印香	Po k'o yin hsiang		印篆	yin and chuan
	Retired Gentleman of the Central Studio, (see Chung-chai chü-shih)		禪林象器篆	Zenrin zōkisen
	Seal, (see yin)		印香	yin-hsiang
	Seal-character, (see chuan)		方干	Fang Kan
山梨	Shan-li		適圉叢書	Shih-yüan ts'ung-shu
沈立	Shen Li		王應麟	Wang Ying-lin
十二枝	Shih erh chih			

五夜香刻   Wu yeh hsiang k'o

印   yin

豫章   Yü-chang

熊朋來   Hsiung P'eng-lai

至治   Chih-chih (Reign)

百刻香印   pai-k'o hsiang-yin

吳正仲   Wu Cheng-chung

一炷香的時候   i-chu hsiang te shih-hou

棒香   pang hsiang

INDEX

50

www.ingramcontent.com/pod-product-compliance
Lightning Source LLC
Chambersburg PA
CBHW081333190326
41458CB00018B/5981